Neanderthal Man

Neanderthal Man

In Search of Lost Genomes

Svante Pääbo

BASIC BOOKS
A Member of the Perseus Books Group
New York

Books published by Basic Books are available at special discounts for bulk purchases in
the United States by corporations, institutions, and other organizations. For more information, please contact the Special Markets Department at the Perseus Books Group, 2300
Chestnut Street, Suite 200, Philadelphia, PA 19103, or call (800) 810-4145, ext. 5000, or
e-mail special.markets@perseusbooks.com.

Designed by Jack Lenzo

Library of Congress Cataloging-in-Publication Data

Pääbo, Svante.
  Neanderthal man : in search of lost genomes / Svante Paabo.
     pages cm
  Includes bibliographical references and index.
   ISBN 978-0-465-02083-6 (hardback) -- ISBN 978-0-465-08068-7 (e-book) 1. Neander-
thals. 2. Human population genetics. 3. Genome analysis. I. Title.
  GN285.P33 2014
  569.9'86--dc23
                            2013041877

10 9 8 7 6 5 4 3 2 1

To Linda, Rune, and Freja

Contents

Preface

The idea to write this book was first suggested to me by John Brockman. Without his initiative and encouragement, I would never have taken the time to write a manuscript much longer than the short scientific articles I am used to authoring. Once I got started, however, I enjoyed the process. Thank you for making this happen!

Many people have helped me by reading the text and suggesting improvements. First of all I thank my wife, Linda Vigilant, who in addition was always supportive of the endeavor, even if it meant me being away from the family. Sarah Lippincot, Carol Rowney, Christine Arden, and, above all, Tom Kelleher at Basic Books were excellent editors. I hope I have learned from them. Carl Hannestad, Kerstin Lexander, Viola Mittag, and others read parts or all of the text and gave helpful suggestions. Souken Danjo provided hospitality in Saikouji for some of the time I needed to withdraw from the world.

I recount events as I remember them. But I suspect that I may have mixed up or conflated a few specifics here and there—for example, regarding various meetings in and trips to Berlin, to 454 Life Sciences, and so on. Obviously, too, I recount events from my own subjective perspective, trying to give credit (and its opposite) where in my opinion it is due. I am aware that this perspective is not the only way one can view such events. In order not to burden the text with too many names and details, I have refrained from mentioning many persons who were nevertheless important. I apologize to everyone who feels unduly ignored!

Chapter 1
Neanderthal *ex Machina*

Late one night in 1996, just as I had dozed off in bed, my phone rang. The caller was Matthias Krings, a graduate student in my laboratory at the Zoological Institute of the University of Munich. All he said was, "It's not human."

"I'm coming," I mumbled, threw on some clothes, and drove across town to the lab. That afternoon, Matthias had started our DNA sequencing machines, feeding them fragments of DNA he had extracted and amplified from a small piece of a Neanderthal arm bone held at the Rheinisches Landesmuseum in Bonn. Years of mostly disappointing results had taught me to keep my expectation low. In all probability, whatever we had extracted was bacterial or human DNA that had infiltrated the bone sometime in the 140 years since it had been unearthed. But on the phone, Matthias had sounded excited. Could he have retrieved genetic material from a Neanderthal? It seemed too much to hope for.

In the lab, I found Matthias along with Ralf Schmitz, a young archaeologist who had helped us get permission to remove the small section of arm bone from the Neanderthal fossil stored in Bonn. They could hardly control their delight as they showed me the string of A's, C's, G's, and T's coming out of one of the sequencers. Neither they nor I had ever seen anything like it before.

What to the uninitiated may seem a random sequence of four letters is in fact shorthand for the chemical structure of DNA, the genetic material stored in almost every cell in the body. The two strands of the famous double helix of DNA are made up of units containing the nucleotides adenine, thymine, guanine, and cytosine, abbreviated A, T, G, and C. The order in which these nucleotides occur makes up the genetic information necessary to form our body and support its functions. The particular piece of DNA we were looking at was part of the mitochondrial genome—mtDNA, for short—that is transmitted in the egg cells of all mothers to their children. Several hundred copies of it are stored in the mitochondria, tiny structures in the cells, and it specifies information necessary for these structures to fulfill their function of producing energy. Each of us carries only one type of mtDNA, which comprises a mere 0.0005 percent of our genome. Since we carry in each cell many thousands of copies of just the one type, it is particularly easy to study, unlike the rest of our DNA—a mere two copies of which are stored in the cell nucleus, one from our mother and one from our father. By 1996, mtDNA sequences had been studied in thousands of humans from around the world. These sequences would typically be compared to the first determined human mtDNA sequence, and this common reference sequence, in turn, could be used to compile a list of which differences were seen at which positions. What excited us was that the sequence we had determined from the Neanderthal bone contained changes that had not been seen in any of those thousands of humans. I could hardly believe that what we were looking at was real.

As I always am when faced with an exciting or unexpected result, I was soon plagued by doubts. I looked for any possibility that what we saw could be wrong. Perhaps someone had used glue produced from cow hide to treat the bones at some point, and we were seeing mtDNA from a cow. No: we immediately checked cow mtDNA (which others had already sequenced) and found that it was very different. This new mtDNA sequence was clearly close to the human sequences, yet it was slightly different from all of them. I began to believe that this was, indeed, the first piece of DNA ever extracted and sequenced from an extinct form of human.

We opened a bottle of champagne kept in a fridge in the lab's coffee room. We knew that, if what we were seeing was really Neanderthal DNA, enormous possibilities had opened up. It might one day be possible to compare whole genes, or any specific gene, in Neanderthals to the corresponding genes in people alive today. As I walked back home through a dark and quiet Munich (I'd had too much champagne to drive), I could hardly believe what had happened. Back in bed, I couldn't sleep. I kept thinking about Neanderthals, and about the specimen whose mtDNA it seemed we had just captured.

In 1856, three years before the publication of Darwin's *The Origin of Species,* workers clearing out a small cave in a quarry in Neander Valley, about seven miles east of Düsseldorf, uncovered the top of a skull and some bones they thought had come from a bear. But within a few years the remains were identified as those of an extinct, perhaps ancestral, form of human. This was the first time that such remains had been described, and the finding shook the world of naturalists. Over the years, research has continued on those bones and many more like them since found, seeking to discern who the Neanderthals were, how they lived, why they disappeared some 30,000 years ago, how our modern ancestors interacted with them over thousands of years of coexistence in Europe, and whether they were friend or foe, our forebears, or simply our long-lost cousins (see Figure 1.1). Tantalizing hints of behaviors familiar to us, such as care of the injured, ritualistic burial, and maybe even the production of music, emerged from archaeological sites, telling us that the Neanderthals were much more like us than is any living ape. How alike? Whether they could speak, whether they were a dead-end branch of the hominin family tree, or whether some of their genes are hidden in us today are all questions that have become an integral part of paleoanthropology, the academic discipline that can be said to have started with the discovery of those bones in Neander Valley, from which we now seemed able to extract genetic information.

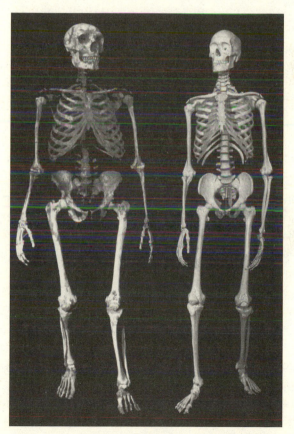

FIGURE 1.1. A reconstructed Neanderthal skeleton (left) and a present-day human skeleton (right). Credit: Ken Mowbray, Blaine Maley, Ian Tattersall, Gary Sawyer, American Museum of Natural History.

As interesting as these questions were in themselves, it seemed to me that the Neanderthal bone fragment held the promise of an even larger prize. Neanderthals are the closest extinct relative of contemporary humans. If we could study their DNA, we would undoubtedly find that their genes were very similar to ours. Some years earlier, my group had sequenced a large number of DNA fragments from the chimpanzee genome and had shown that in DNA sequences we shared with the chimpanzees, only a bit over 1 percent of the nucleotides differed. Clearly, the Neanderthals must be much closer to us than that. But—and this is what was immensely exciting—among the few differences one would expect to find in the Neanderthal genome, there must be those that set us apart from all earlier forms of human forerunners: not just from the Neanderthals but also from Turkana Boy, who lived some 1.6 million years ago; Lucy, some 3.2 million years ago; and Peking Man, more than half a million years ago. Those few differences must form the biological foundations of the radically new direction our lineage took with the emergence of modern humans: the advent of rapidly developing technology, of art in a form we today immediately recognize as art, and maybe of language and culture as we now know it. If we could study Neanderthal DNA, all this would be within our grasp. Wrapped in such dreams (or delusions of grandeur), I finally drifted off to sleep as the sun rose.

The next day Matthias and I both arrived late at the lab. After checking the DNA sequence from the night before to make sure we had not made any mistakes, we sat down and planned what to do next. It was one thing to get the sequence of one little piece of mtDNA that looked interesting from the Neanderthal fossil, but it would be quite another to convince ourselves, let alone the rest of the world, that it was mtDNA from an individual who lived (in this particular case) some 40,000 years ago. My own work over the previous twelve years made our next step fairly clear. First, we needed to repeat the experiment—not just the last step but all the steps, beginning with a new piece of the bone in order to show that the sequence we had obtained was not some fluke derived from a badly damaged and modified modern mtDNA molecule in the bone. Second, we needed to extend the sequence of mtDNA we had obtained by retrieving overlapping DNA fragments from the bone extract. This would enable us to reconstruct a longer mtDNA sequence, with which we could begin to estimate just how different the mtDNA of Neanderthals was from that of humans today. And then a third step was necessary. I myself had often suggested that extraordinary claims about DNA sequences from ancient bones require extraordinary evidence—namely, repetition of the results in another lab, an unusual step in a typically competitive scientific field. The claim that we had retrieved Neanderthal DNA would certainly be considered extraordinary. To exclude unknown sources of error in our lab, we needed to share some of the precious bone material with an independent lab and hope that it could manage to repeat our result. All of this I discussed with Matthias and Ralf. We laid out plans for the work and swore one another to absolute secrecy outside our research groups. We wanted no attention until we were sure that what we had was the real thing.

Matthias got to work at once. Having spent almost three years on mostly fruitless attempts to extract DNA from Egyptian mummies, he was energized by the prospect of success. Ralf seemed frustrated over having to return to Bonn, where he could do nothing but eagerly await word of our results. I tried to concentrate on my other projects, but it was hard to take my mind off what Matthias was doing.

What Matthias needed to do was not all that easy. We were dealing, after all, with something other than the intact and pristine DNA that comes from a blood sample drawn from a living person. The neat and tidy double-stranded, helical DNA molecule in the textbooks—with its nucleotides A, T, G, and C, attached in complementary pairs (adenine with thymine, guanine with cytosine) to the two sugar-phosphate backbones—is not a static chemical structure when stored in the nuclei and mitochondria of our cells. Rather, DNA continually suffers chemical damage, which is recognized and repaired by intricate mechanisms. In addition, DNA molecules are extremely long. Each of the twenty-three pairs of chromosomes in the nucleus comprises one enormous DNA molecule; the total length of one set of twenty-three chromosomes adds up to about 3.2 billion nucleotide pairs. Since the nucleus has two copies of the genome (each copy stored on one set of twenty-three chromosomes, of which we inherit one from our mother and one from our father), it contains about 6.4 billion nucleotide pairs. By comparison, the mitochondrial DNA is tiny, with a little over 16,500 nucleotide pairs; but given that the mtDNA we had was ancient, the challenge involved in sequencing it was great.

The most common type of damage that occurs spontaneously in DNA molecules, whether nuclear DNA or mtDNA, is the loss of a chemical component—an amino group—from the cytosine nucleotide (C), turning it into a nucleotide that does not naturally occur in DNA called uracil, abbreviated U. There are enzyme systems in the cells that remove these U's and replace them with the correct nucleotide, C. The discarded U's end up as cellular garbage, and from analyses of damaged nucleotides excreted in our urine it has been calculated that about ten thousand C's per cell morph into U's each day, only to be removed and then replaced. And this is just one of several types of chemical assaults our genome suffers. For example, nucleotides are lost, creating empty sites that quickly lead to breakage of the strands in the DNA molecules. Working against this are enzymes that fill in such missing nucleotides before a break can occur. If a break does occur, other enzymes join the DNA molecules back together. In fact, the genomes in our cells would not remain intact for even an hour if these repair systems were not there to maintain them.

These repair systems, of course, require energy to work. When we die, we stop breathing; the cells in our body then run out of oxygen, and as a consequence their energy runs out. This stops the repair of DNA, and various sorts of damage rapidly accumulate. In addition to the spontaneous chemical damage that continually occurs in living cells, there are forms of damage that occur after death, once the cells start to decompose. One of the crucial functions of living cells is to maintain compartments where enzymes and other substances are kept separate from one another. Some of these compartments contain enzymes that can cut DNA strands and are necessary for certain types of repair. Other compartments contain enzymes that break down DNA from various microorganisms that the cell may encounter and engulf. Once an organism dies and runs out of energy, the compartment membranes deteriorate, and these enzymes leak out and begin degrading DNA in an uncontrolled way. Within hours and sometimes days after death, the DNA strands in our body are cut into smaller and smaller pieces, while various other forms of damage accumulate. At the same time, bacteria that live in our intestines and lungs start growing uncontrollably when our body fails to maintain the barriers that normally contain them. Together these processes will eventually dissolve the genetic information stored in our DNA—the information that once allowed our body to form, be maintained, and function. When that process is complete, the last trace of our biological uniqueness is gone. In a sense, our physical death is then complete.

However, nearly each one of the trillions of cells in our body contains the entire complement of our DNA. Thus, even if the DNA in some cells in some remote corner of our body evades complete decomposition, some trace of our genetic makeup will survive. For example, the enzymatic processes that degrade and modify the DNA require water to work. If some parts of our body dry out before the degradation of the DNA has run its course, these processes will stop, and fragments of our DNA may survive for a longer time. This occurs, for example, when a body is deposited in a dry place where it becomes mummified. Such whole-body desiccation may occur accidentally, owing to the environment in which the body happens to end up, or it may be deliberately practiced. As is commonly known, ritual mummification of the dead was often performed in ancient Egypt, where the bodies of hundreds of thousands of people who lived between about 5,000 and 1,500 years ago were mummified to provide *post mortem* abodes for their souls.

Even when mummification does not occur, some parts of the body, such as the bones and teeth, may survive long after burial. These hard tissues contain cells, responsible for tasks such as forming new bone when a bone is broken, that are embedded in microscopically small holes. When these bone cells die, their DNA may leak out and become bound to the mineral component of the bone, where it may be shielded from further enzymatic attack. Thus, with luck, some DNA may survive the onslaught of degradation and damage that occurs in the immediate aftermath of death.

But even when some DNA survives the bodily chaos that follows death, other processes will continue to degrade our genetic information, albeit at a slower rate. For example, the continuous flow of background radiation that hits Earth from space creates reactive molecules that modify and break DNA. Furthermore, processes that require water—such as the loss of amino groups from the nucleotide C, resulting in U nucleotides—continue even when DNA is preserved under relatively dry conditions. This is because DNA has such an affinity for water that even in dry environments water molecules are almost always bound to the grooves between the two DNA strands, allowing spontaneous water-dependent chemical reactions to occur. The loss of the amino group—or deamination—of the nucleotide C is one of the fastest of these processes, and it will destabilize the DNA so that its strands eventually break. These and other processes, most still unknown, keep chipping away at our DNA even when it has survived the havoc that death itself causes in our cells. Although the rate of despoliation will depend on many circumstances such as temperature, acidity, and more, it is clear that even under favorable conditions, ultimately even the last pieces of information from the genetic program that made a person possible will eventually be destroyed. It seemed that in the Neanderthal bone my colleagues and I had analyzed, all these processes had not yet fully completed their destructive task after 40,000 years.

Matthias had retrieved the sequence of a piece of mtDNA 61 nucleotides long. To do so, he had to make many copies of this piece of DNA, which involved a process called the polymerase chain reaction (PCR). In his attempt to confirm our findings, he started by repeating his PCR experiment exactly as he had done it the first time. This experiment involves using two short synthetic pieces of DNA called primers, which were designed to bind to two places in the mtDNA, 61 nucleotide pairs apart. These primers are mixed with a tiny amount of the DNA extracted from the bone and an enzyme called DNA polymerase that can synthesize new DNA strands starting and ending with the primers. The mixture is heated to allow the two DNA strands to come apart, so that the primers can find and bind to their target sequences when the mixture is cooled via pairings of A's with T's and G's with C's. The enzyme will then use the primers bound to the DNA strands as starting points to synthesize two new strands, duplicating the two original strands from the bone, so that these two original strands become four strands. This amplification process is repeated to produce eight strands, then again to produce sixteen, then thirty-two, and so on for a total of thirty or forty rounds of duplication.

The PCR—a simple yet elegant technique invented by the maverick scientist Kary Mullis in 1983—is extremely powerful. From a single DNA fragment, one can, in principle, obtain about a trillion copies after forty cycles. This is what made our work possible, so in my opinion Mullis certainly deserved the Nobel Prize in chemistry that he was awarded in 1993. However, the PCR's exquisite sensitivity also made our work difficult. In an extract from an ancient bone—which may contain very few surviving ancient DNA molecules, or none at all—there might be one or more molecules of modern human DNA that have contaminated the experiment: from the chemicals used, from the lab's plasticware, or from airborne dust. Dust particles in rooms where humans live or work are, to a large extent, human skin fragments, which contain cells full of DNA. Alternatively, human DNA might have contaminated the sample when a person handled it—say, in a museum or at an excavation. It was with these concerns in mind that we chose to study the sequence of the most variable part of the Neanderthal mtDNA. Since many humans differ from one another in that particular section, we could at least tell whether more than one human had contributed DNA to our experiment and thus be warned that something was amiss. This is why we were so excited about finding a DNA sequence with changes never before seen in any human; had the sequence looked like that of a living human, we would have had no way of determining whether it meant, on the one hand, that the Neanderthal was indeed identical in mtDNA to some people today or, on the other, that we were just looking at a modern mtDNA fragment that had made its way into our experiments from some insidious source such as a speck of dust.

By this time, I was all too acquainted with the fact of contamination. I had been working for more than twelve years on the extraction and analysis of ancient DNA from such extinct mammals as cave bears, woolly mammoths, and ground sloths. After a string of frustrating results (I detected human mtDNA in almost all the animal bones I analyzed with the PCR), I spent a great deal of time thinking about, and devising, ways to minimize contamination. Thus, Matthias performed all extractions and other experiments, right up to the first temperature cycle of the PCR, in a small lab that was kept meticulously clean and absolutely separate from the rest of our laboratory. Once the ancient DNA, the primers, and the other components necessary for the PCR had been placed together in a tube, the tube was sealed, and the temperature cycles and all subsequent experiments were performed in the regular laboratory. In the clean lab, all surfaces were washed with bleach once a week, and every night the lab was irradiated with UV light to destroy any dust-borne DNA. Matthias entered the clean lab through an antechamber, where he and others working there donned protective gowns, face shields, hairnets, and sterile gloves. All reagents and instruments were delivered directly to the clean lab; nothing was allowed into it from other parts of the institute. Matthias and his colleagues had to start their workday in the clean lab instead of in other parts of our laboratory, where large amounts of DNA were being analyzed. Once they had entered any such lab, they were banned from the clean lab for the rest of the day. To put it mildly, I was paranoid about DNA contamination, and I felt I had good reason to be.

Even so, in Matthias's initial experiments we had seen evidence of some human contamination of the bone. After using the PCR to amplify a piece of mtDNA from the bone, he had cloned the resulting batch of supposedly identical DNA copies in bacteria. He did this in order to see whether more than one type of mtDNA sequence might exist among the cloned molecules: each bacterium will take up just one 61-nucleotide molecule joined to a carrier molecule called a plasmid and grow up to a clone of millions of bacteria where each carries copies of the 61-nucleotide molecule the first bacterium took up, so we were able to get an overview of whatever different DNA sequences existed in the population of molecules by sequencing a number of the clones. In Matthias's earliest experiment, we saw seventeen cloned molecules that were similar or identical to one another while different from the two thousand–plus modern human mtDNAs we were using for comparison—but we also saw one that was identical to a sequence seen in some humans today. This clearly showed the presence of contamination, perhaps from museum curators or others who had handled the bone over the 140 years since its discovery.

So the first thing Matthias did in his attempt to reproduce our original result was to repeat the PCR and the cloning. This time he found ten clones with the unique sequence that had excited us so much and two that looked like they could have come from any modern person. He then went back to the bone and made another extract, did the PCR and cloning again, and got ten clones of the interesting type and four that looked like mtDNA from present-day humans. Now we were satisfied: our original result had passed the first tests—we could repeat them and see the same unusual DNA sequence each time.

Matthias next started to "walk along" the mtDNA, using other primers designed to amplify fragments that overlapped a part of the first fragment but extended further into other regions of the mtDNA (see Figure 1.2). Once again, we observed that some of the sequences of these fragments had nucleotide changes never seen in contemporary humans. Over the next few months, Matthias amplified thirteen different DNA fragments of different sizes, each at least twice. The interpretation of the sequences was complicated by the fact that any one DNA molecule can carry mutations that can be due to various causes: ancient chemical modifications, sequencing errors, or even rare but natural variation that may exist among the mtDNA molecules found within a cell of an individual. Therefore, we used a strategy I had worked out previously for ancient animal DNA (again, see Figure 1.2). For each position in each experiment, we took as authentic the so-called consensus nucleotide—that is, the nucleotide (A, T, G, or C) carried in that position by most of the molecules we examined. We also required that each position be identical in at least two independent experiments, since a PCR might, in an extreme case, have started from a single DNA strand—in which case all the clones, due to some error in the first PCR cycle or to some chemical modification in that particular DNA

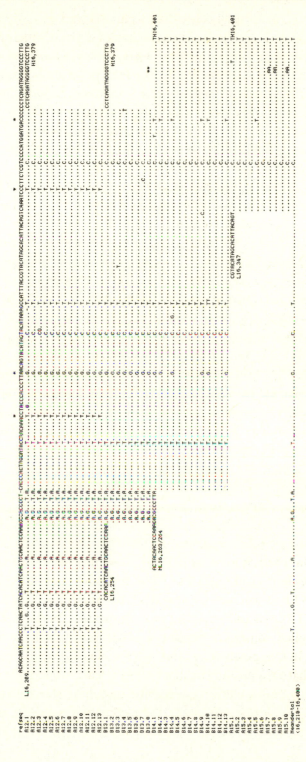

FIGURE 1.2. Reconstruction of a piece of mtDNA from the Neanderthal from Neander Valley. Above, a modern reference sequence is shown. Each line below represents one cloned molecule amplified from the Neanderthal type specimen. Where these sequences are identical to the reference sequence, I have placed a dot; where they differ from the nucleotide, I have written them out. In the bottom line is the reconstructed Neanderthal nucleotide sequence. At each position, we require that a change from the reference sequence is seen in a majority of clones and in at least two independent PCR experiments (either the ones shown or others). From Matthias Krings et al., "Neandertal DNA sequences and the origin of modern humans," *Cell* 90, 19–30 (1997).

strand, would carry the same nucleotide in the same position. If two PCRs differed with regard to even a single position, we repeated the PCR a third time, to see which of the two nucleotides was reproducible. Matthias eventually used 123 cloned DNA molecules to puzzle together a sequence of 379 nucleotides of the most variable part of the mtDNA. According to the criteria we had settled on, this was the DNA sequence that this Neanderthal carried when he or she was alive. Once we had this longer sequence, we could begin the exciting work of comparing it with the variations seen in present-day humans.

At this point, we compared our 379-nucleotide Neanderthal mtDNA sequence to the corresponding mtDNA sequences from 2,051 present-day humans from all around the world. On average, twenty-eight of the positions differed between the Neanderthal and a contemporary person, whereas people alive today carry an average of only seven differences from one another. The Neanderthal mtDNA was four times as different.

We then looked for any indication that the Neanderthal mtDNA was more like the mtDNA found in modern Europeans. One might well expect to find this, since Neanderthals evolved and lived in Europe and western Asia; indeed, some paleontologists believe that Neanderthals are among the ancestors of today's Europeans. We compared the Neanderthal mtDNA with that of 510 Europeans and discovered that it carried, on average, twenty-eight differences. We then compared it with mtDNA from 478 Africans and 494 Asians. The average number of differences from the mtDNAs of these people was also twenty-eight. This meant that, on average, European mtDNAs were no more similar to the Neanderthal mtDNA than were mtDNAs from modern-day Africans and Asians. But maybe Neanderthal mtDNAs were similar to mtDNAs found in just some Europeans, as one would expect if Neanderthals had contributed some mtDNA to Europeans. We checked this and found that the Europeans in our sample whose mtDNAs were most like that of the Neanderthal showed twenty-three differences; the Africans and Asians closest to the Neanderthal carried twenty-two and twenty-three, respectively. In short, we observed not only that the Neanderthal mtDNA seemed very different from the mtDNAs of modern humans worldwide but also that there was no indication of any special relationship between the Neanderthal mtDNA and any subset of European mtDNAs alive today.

However, just counting differences is not enough to reconstruct the evolutionary history of a piece of DNA. The differences found between DNA sequences represent mutations that occurred in the past. But some types of mutations are more frequent than others, and some positions in DNA sequences are more prone to mutate than others. At such positions, more than one mutation—especially the types that happen more frequently—may have occurred in the history of a DNA sequence. Therefore, to estimate the history of this particular piece of mtDNA, we needed to apply models for how we believed it had mutated and evolved, bearing in mind that certain positions might have mutated more than once, thus obscuring previous mutations. The result of such a reconstruction is depicted as a tree, in which a DNA sequence on the tip of a branch links back to a common ancestral DNA sequence. These ancestral sequences are depicted as the points where branches join on the tree (see Figure 1.3). When we did such a tree reconstruction, we saw that the mtDNAs of all humans alive today trace their ancestry back to one common mtDNA ancestor.

This finding, which was already known from Allan Wilson's work in the 1980s,[1] is in fact expected for mtDNA, since each of us carries only a single type and cannot exchange pieces of it with other mtDNA molecules in the population. Since mtDNA is passed on only by mothers, the mtDNA

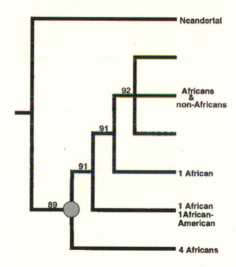

FIGURE 1.3. A mtDNA tree, illustrating how the mtDNAs of people alive today trace their ancestry back to a common mtDNA ancestor (the so-called Mitochondrial Eve, indicated by a circle) who existed more recently than the mtDNA ancestor shared with the Neanderthal. Nucleotide differences are used to infer branching order, and the numbers refer to the statistical support for the branching order shown. Modified from Matthias Krings et al., "Neandertal DNA sequences and the origin of modern humans," *Cell* 90, 19–30 (1997).

lineage of a woman will die out if she has no female descendants—so in each generation some mtDNA lineages vanish. Therefore, there must once have existed a woman—the so-called Mitochondrial Eve—who carried an mtDNA lineage that would turn out to be the ancestor of all human mtDNAs today, simply because all other lineages since that time have been lost, purely by chance.

According to our models, however, the Neanderthal mtDNA did not trace back to this Mitochondrial Eve but went further back before it shared an ancestor with the mtDNAs of humans alive today. This finding was immensely exciting. It proved beyond any doubt that we had recovered a piece of Neanderthal DNA—and it showed, at least with respect to their mtDNA, that the Neanderthals were profoundly different from us.

My colleagues and I also used the models to estimate how long ago the Neanderthal mtDNA shared an ancestor with current human mtDNAs. The number of differences between the two types of mtDNA is an indication of how long they have been transmitted through generations independently of each other. The mutation rates of widely separated species—mice and monkeys, say—will differ, but among closely related species—such as humans, Neanderthals, and the great apes—they have been constant enough to allow scientists to estimate, based on the differences observed, when two DNA sequences last shared an ancestor. Using the models for how fast different types of mutations occur in mtDNA, we estimated that the mtDNA ancestor common to all humans alive today, the Mitochondrial Eve, lived between 100,000 and 200,000 years ago, as Allan Wilson and his team had found. However, the ancestor that the Neanderthal mtDNA shared with human mtDNAs lived about 500,000 years ago; that is, she was three or four times as ancient as the Mitochondrial Eve from whom all present-day human mtDNAs are descended.

This was amazing stuff. I was now fully convinced that we had recovered Neanderthal DNA and that it was very different from the DNA of modern humans. However, before publishing our findings, we needed to overcome the last hurdle: we needed to find an independent laboratory that could repeat what we had done. Such a lab would not need to determine the entire 379-nucleotide mtDNA sequence, but it *would* have to retrieve one of the regions that carried one or more substitutions setting Neanderthals apart from humans today. This would show that the DNA sequence we had determined really existed in the bone and wasn't some strange and unknown sequence perhaps floating around in our laboratory. But to whom could we turn? This was a delicate issue.

Although many labs would undoubtedly want to participate in such a potentially high-profile project, there was the risk that if we chose one that had not worked as intensely as we had on minimizing contamination and addressing all the other problems associated with ancient DNA, it might fail to successfully extract and amplify a relevant sequence. If that happened, our results would be deemed irreproducible and thus unpublishable. I knew that no one had spent as much time and effort on this sort of work as we had, but we eventually settled on the laboratory of Mark Stoneking, a population geneticist at Penn State University. Mark had been a graduate student and then a postdoc with Allan Wilson at Berkeley, and I knew him from my time spent as a postdoc there in the late 1980s. He was one of the people behind the discovery of the Mitochondrial Eve and one of the architects of the out-of-Africa hypothesis of modern human origins—the idea that modern humans originated in Africa some 100,000 to 200,000 years ago, then spread around the world and replaced all earlier forms of humans, such as the Neanderthals in Europe, without admixture. I respected his judgment and integrity and knew him to be an easygoing person. Moreover, one of his graduate students, Anne Stone, had spent the 1992–1993 academic year in our laboratory. Anne, a serious-minded and ambitious scientist, had worked with us on the retrieval of mtDNA from some Native American skeletal remains, so she was familiar with our techniques. I felt that if anyone could repeat what we had done, she could.

I contacted Mark. As expected, he and Anne were excited about trying this, so we parted with one of the last pieces of bone that Ralf had given us. We told Anne and Mark which part of the mtDNA they should try to amplify, so that they would have the best chance of hitting one of the positions in the mtDNA sequence that carried a mutation typical of our Neanderthal sequence. But we sent them no primers or other reagents, just a piece of bone that had been kept in a sealed tube since its trip from Bonn. This precaution minimized any chance that a contaminant might pass from our lab to theirs. We also did not tell them what positions were typical of Neanderthal mtDNA, not because I didn't trust them but because I wanted to be able to say we had done everything we could to avoid even an unconscious bias. In short, Anne would have to synthesize the primers and do everything independently of us without knowing exactly what result we expected. Once we'd sent the bone off by FedEx, we just had to wait.

Generally, these sorts of experiments take longer than expected: a company fails to deliver primers in the promised time, a reagent you test for contamination turns out to have human DNA in it, a technician falls sick the day he is to run the sequencing machine with the crucial sample. We waited for what seemed like an eternity for Anne to call from Pennsylvania. And then one night she did call. The tone of her voice immediately told me she was not happy. She had cloned fifteen amplified DNA molecules from the region of interest, and they all looked like any person's today—in fact, like my own or Anne's mtDNA. This was a crushing defeat. What did it mean? Had we amplified some freak mtDNA? I could not believe that was the case. If it was from some unknown animal, it would not be as close to human mtDNA as it was, yet it could hardly be mtDNA from some unusual human if it was roughly four times as different as all the human mtDNAs that had been studied. There was always the possibility that the sequence we had come up with was created by some chemical modification of the ancient DNA that consistently attacked the same positions in the sequence; however, such a modified mtDNA sequence would be expected to look like a human sequence with extra changes added by this unknown chemical process, rather than like a sequence that had branched off the human lineage in the past. And even then, why wouldn't Anne find the same sequence we did? The only plausible explanation seemed to be that Anne had more contamination in her experiments than we did—so much so that it outnumbered the rare Neanderthal molecules. What could we do? We could hardly go back to Ralf and ask him for another piece of the valuable fossil on the chance that the next experiment would be more successful than the first.

Perhaps, even if Anne's experiments had more contamination than ours did, she could sequence thousands of mtDNA molecules from her piece of bone and thereby find some rare ones that looked like ours. But in the meantime we had done experiments to estimate the number of Neanderthal mtDNA molecules in the Neanderthal bone extracts we had used to start our PCRs. As it turned out, there were about fifty. By comparison, a source of contamination such as a dust particle might contain tens of thousands, or hundreds of thousands, of mtDNA molecules. So any such fishing expedition was very likely to fail.

I discussed this conundrum at length, not only with Matthias but in our weekly lab meetings with the subgroup of my lab working on ancient DNA. Throughout my career I have found these extensive discussions with scientists working in my lab to be very useful; indeed, I think they have been crucial to whatever successes we have had. In such discussions, ideas that would never occur to people focusing solely on their own work are often hatched. Moreover, scientists without a personal stake in a project's outcome provide a reality check, since they are free of the wishful thinking all too common among those who are working on a project they love and on which their scientific future may depend. Often my role in these discussions is to moderate and select the ideas that seem promising enough to pursue.

Once again, our meeting bore fruit, and we came up with a plan. Anne would be asked to make primers that would not be a perfect match to modern DNA. Instead, the final nucleotide on their tips would match a nucleotide uniquely seen in our putative Neanderthal sequence. Such primers would not (or only very weakly) initiate amplification from modern human mtDNA and thus would favor the amplification of Neanderthal-like mtDNAs. We discussed this plan thoroughly—especially the crucial point of whether Anne's effort could be considered an independent replication of our finding if she used information from our sequence to make the primers. Obviously, it would have been more aesthetically pleasing if Anne had been able to come up with the same sequence that we did without any prior knowledge of the sequence. However, we could tell her to synthesize Neanderthal-specific primers that would bracket two other positions that also carried unique nucleotides. And we would not tell her where or how many of those positions there were. If she found the same unique nucleotide changes that we had, then we would all be convinced that such molecules were indeed native to the bone itself. After much further discussion, we agreed that this was a legitimate way forward.

We transmitted the necessary information to Anne; she ordered the new primers; we waited. By now it was mid-December, and Anne had told us earlier that she planned to fly to North Carolina to visit her parents over Christmas. I obviously could not tell her to cancel, much as I wished she would. Finally, after almost two weeks, the phone rang. Anne had sequenced five molecules from her new PCR products. All of them contained the two substitutions we had seen in our Neanderthal sequence—substitutions that are rare or absent in modern humans. This was an enormous relief. I felt we all deserved a Christmas break. We called Ralf in Bonn to relay the good news. As I'd often done during my years in Munich, I celebrated New Year by taking a skiing trip with some wildlife biologists to remote valleys in the Alps on the Austrian border. This time, while skiing up the spectacular valleys, I could not refrain from formulating the paper that would describe the first DNA sequence from a Neanderthal. To me, what we were about to describe was even more spectacular than the steep and snowy landscape surrounding me.

Matthias and I met up again in the lab after Christmas vacations and sat down to write our paper. One major question was where to send it. *Nature,* the British journal, and its American counterpart *Science,* enjoy the most prestige and visibility in the scientific community and in the general media, and either would have been an obvious choice. But they both impose strict length limits on manuscripts, and I wanted to explain all the details of what we had done—not only to convince the world that we had the real thing but also to promote our painstaking methods of extracting and analyzing ancient DNA. In addition, I had become disenchanted with both journals because of their tendency to publish flashy ancient DNA results that did not meet the scientific criteria our group considered necessary. They often seemed more interested in publishing papers that would give them coverage in the *New York Times* and other major media outlets than in making sure the results were sound and likely to hold up.

I discussed all this with Tomas Lindahl, a Swedish-born scientist at the Imperial Cancer Research Fund Laboratory in London. Tomas, a preeminent expert in DNA damage, is soft-spoken, yet not one to shy away from controversy when he knows he's right. He has been something of a mentor to me since 1985, when I spent six weeks in his laboratory studying chemical damage in ancient DNA. Tomas suggested we send the paper to *Cell,* a highly respected and influential journal that specializes in molecular and cell biology. Publication there would send a signal to the community that the sequencing of ancient DNA was solid molecular biology and not just about the production of sexy but questionable results; moreover, *Cell* allowed long articles. Tomas called its celebrated editor, Benjamin Lewin, to gauge his interest, since such a manuscript was somewhat outside *Cell*'s usual scope. Lewin told us to submit it and he would send it out for the usual peer review. This was great news. We now had sufficient space in which to describe all our experiments and present all the arguments for why we were convinced we had genuine Neanderthal DNA.

Today, I still consider this paper to be one of my best. In addition to describing the painstaking way in which we had reconstructed the mtDNA sequence and why we considered it genuine, it laid out the evidence that our mtDNA sequence fell outside the range of variation seen today and the implication that Neanderthals had not contributed mtDNA to modern humans. These conclusions were compatible with the out-of-Africa model of human evolution that Allan Wilson, Mark Stoneking, and others had proposed. As my colleagues and I said in our paper: "The Neandertal mtDNA sequence thus supports a scenario in which modern humans arose recently in Africa as a distinct species and replaced Neandertals with little or no interbreeding."

We also tried to describe all the caveats we could think of. In particular, we pointed out that mtDNA offers only a limited view of the genetic history of a species. Because it is transmitted only from mothers to offspring, it reflects exclusively the female side of history. Therefore, if Neanderthals interbred with modern humans, we would detect it only if females crossed over between the two groups. This need not have been the case. In more recent human history, when human groups that differ in social status have met and interacted, they almost always had sex with one another and produced offspring. But this generally occurred in a biased way with respect to what males and females do: in other words, the partner from the socially dominant group was most often male, and the offspring of these unions tended to remain in their mother's group. Of course, we do not know if such a pattern was typical of modern humans when they came to Europe and met Neanderthals some 35,000 years ago. And we do not even know if modern humans were socially dominant in any sense that would be comparable to what we see among human groups today. But it is clear that looking only at the female side of inheritance tells us only half the story of what happened.

Another, even more important limitation of mtDNA stems from the way it is inherited. As noted, an individual's mtDNA does not exchange bits and pieces with another individual's mtDNA. Furthermore, if a woman has only sons, her mtDNA will become extinct. Because chance plays such a strong role in the history of mtDNA, even if some had passed from Neanderthals to early modern humans in Europe at some point between 35,000 and 30,000 years ago, it may well have disappeared. This limitation does not exist for the chromosomes in the cell nucleus: recall that they exist in pairs in every individual, with one chromosome in the pair coming from the mother and the other coming from the father. When sperm or egg cells are formed in an individual, the chromosomes break and rejoin in an intricate dance that results in pieces being exchanged between them. Therefore, if we are able to study several parts of an individual's nuclear genome, we would get several different versions of the genetic history of a group.

For example, even if, in some parts, the variants perhaps contributed by Neanderthals were lost, this would probably not be the case for all parts. Therefore, by looking at many parts of the nuclear genome, one arrives at a picture of human history that is much less influenced by chance. For this reason, we concluded in our paper that our results "do not rule out the possibility that Neandertals contributed other genes to modern humans." However, given the evidence at hand, we clearly favored the out-of-Africa hypothesis.

Our paper was peer-reviewed and accepted for publication by *Cell* after only minor revisions. As is typical for all top journals, the editors at *Cell* insisted that we not talk about our results before publication in the July 11 issue.[2] They prepared a press release, and I flew to the press conference they organized in London for the day of the publication. It was my first press conference and the first time I'd ever found myself at the center of such intense media attention. Much to my surprise, I enjoyed trying to get across the essence of our work, doing my best to describe both our conclusions and the caveats involved. It was not that easy, because our data had direct implications for a pitched battle that had been raging in the field of anthropology for over ten years.

This battle had been initiated by the out-of-Africa hypothesis, which Allan Wilson and his colleagues had proposed based largely on the patterns of present-day human mtDNA variation. Initially, the idea had been met with ridicule and hostility by the paleontological community. Almost all paleontologists at the time subscribed to the so-called multiregional model for the origin of modern humans—holding that modern humans evolved on several continents, more or less independently, from *Homo erectus*. They saw a deep history dividing current groups of humans: the ancestors of current Europeans, for instance, were thought to be the Neanderthals and perhaps earlier European hominins; the ancestors of current Asians were thought to be other archaic forms in Asia, going back to Peking Man. However, a growing number of respected paleontologists, foremost among them Chris Stringer at the Natural History Museum in London, now viewed the out-of-Africa model of modern human origins as the best fit to both the fossil record and the archaeological evidence. Chris had been invited by *Cell* to the press conference, where he announced that our retrieval of Neanderthal DNA was to paleontology what the lunar landing had been to space exploration. I was of course pleased, though not surprised, by his praise. I was even more pleased when the "other side," the multiregionalists, had good things to say at least about the technical aspects of our work—particularly when the most vociferous and pugnacious of them, Milford Wolpoff, of the University of Michigan, declared in a commentary in *Science* that "if anyone would be able to do this, it would be Svante."

All in all, I was stunned by the attention our paper received. It was reported on the first page of many major newspapers and on radio and TV news shows worldwide. In the week after the paper appeared, I spent most of my time on the phone with journalists. I had worked on ancient DNA since 1984 and had gradually realized that it must in principle be possible to recover Neanderthal DNA. And nine months had now passed since Matthias called and awakened me to say he saw a DNA sequence that did not look human come out of one of our sequencing machines. So I'd had time to get used to the idea and, unlike most of the rest of the world, was not shaken by our achievement. Once the media frenzy had died down, though, I felt the need for some perspective. I wanted to reflect on the years that had led up to this discovery and to think about where I would go next.

Chapter 2
Mummies and Molecules

It did not begin with Neanderthals, but with ancient Egyptian mummies. Ever since my mother took me to Egypt when I was thirteen, I had been fascinated with its ancient history. But when I started to pursue this study in earnest, at the University of Uppsala in my native Sweden, it became increasingly clear that my fascination with Pharaohs, pyramids, and mummies was the romantic dream of an adolescent. I did my homework; I memorized the hieroglyphs and the historical facts; I even worked two consecutive summers cataloging pottery shards and other artifacts at the Mediterranean Museum in Stockholm, which might well have become my future workplace, were I to become an Egyptologist in Sweden. I found that the same people did very much the same things the second summer as they had the first summer. Moreover, they went to lunch at the same time, to the same restaurant, ordered the same meals, discussed the same Egyptological puzzles and academic gossip. In essence, I came to realize that the discipline of Egyptology was moving too slowly for my tastes. It was not the kind of professional life I imagined for myself. I wanted more excitement, and more relevance to the world I saw around me.

This disenchantment threw me into a crisis of sorts. In response, and inspired by my father, who had been an MD and later became a biochemist, I decided to study medicine, with a view to doing basic research. I entered medical school at the University of Uppsala and after a few years surprised myself by how much I enjoyed seeing patients. It seemed to be one of the few professions in which you not only met all sorts of people but could also play a positive role in their lives. This ability to engage with people was an unexpected talent, and after four years of medical studies I had another mini-crisis: Should I become a clinician or move, as I had originally meant to do, into basic research? I opted for the latter, thinking that I could—and most likely would—come back to the hospital after my PhD. I joined the lab of one of the then-hottest scientists in Uppsala, Per Pettersson. Not long before, his group had been the first to clone the genetic sequence of an important class of transplantation antigens, protein molecules that sit on the surface of immune cells and mediate their recognition of viral and bacterial proteins. Not only had Pettersson produced exciting biology insights with relevance to clinical practice, but his lab was one of the few in Uppsala that had mastered the then-novel methods of cloning and manipulating DNA by introducing it into bacteria.

Pettersson asked me to join his group's efforts to study a protein encoded by an adenovirus, a virus that causes diarrhea, cold-like symptoms, and other unpleasant features of our lives. It was thought that this viral protein became bound by the transplantation antigens inside the cell, so that, once transported to the cell surface, it could be recognized by immune-system cells, which would then become active and kill other infected cells in the body. Over the next three years, I and the others working on this protein came to realize that this idea of what the protein did was utterly wrong. We found that rather than becoming a hapless target of the immune system, the viral protein seeks out the transplantation antigens inside the cell, binds to them, and blocks their transport out to the cell surface. Since the infected cell thus ends up having no transplantation antigens on its surface, the immune system cannot recognize that it is infected. This protein camouflages the adenovirus, so to speak. In fact, it leads to the creation of a cell within which the adenovirus can probably survive for a long time, perhaps even as long as the infected person lives. That viruses could foil the immune system of their hosts in this way was a revelation, and our work resulted in a number of high-profile papers in the best journals. Indeed, it turns out that other viruses, too, use similar mechanisms to evade the immune system.

This was my first taste of cutting-edge science, and it was fascinating. It was also the first (but not the last) time I saw that progress in science often entails a painful process of realizing that your ideas and those of your peers are wrong, and an even longer struggle to persuade your closest associates and then the world at large to consider a new idea.

But somehow, in the midst of all the biological excitement, I could not quite shake off my romantic fascination with ancient Egypt. Whenever I had time, I went to lectures at the Institute of Egyptology, and I continued to take classes in Coptic, the language of pharaonic Egypt as spoken during the Christian era. I befriended Rostislav Holthoer, a jovial Finnish Egyptologist with an immense capacity for friendships across social, political, and cultural boundaries. During long dinners and evenings at Rosti's home in Uppsala in the late 1970s and early '80s, I often complained that I loved Egyptology but saw little future in it, while I also loved molecular biology, with its apparently boundless promise of advances in the welfare of humankind. I was torn between two equally alluring career paths—a conundrum no less painful because it was doubtless viewed without much sympathy as the fretting of a young man faced with nothing but good choices.

But Rosti was patient with me. He listened when I explained how scientists could now take DNA from any organism (be it a fungus, a virus, a plant, an animal, or a human), join it to a plasmid (a carrier molecule made of DNA from a bacterial virus), and introduce the plasmid into bacteria, where it would replicate along with its host, making hundreds or thousands of copies of the foreign DNA. I explained how we could then determine the sequence of the foreign DNA's four nucleotides and find differences in the sequences between the DNAs of two individuals or two species. The more similar two sequences were—that is, the fewer the number of differences between them—the more closely related they were. In fact, from the number of shared mutations we could infer not only how the particular sequences had evolved from common ancestral DNA sequences over thousands and millions of years but also approximately when those ancestral DNA sequences had existed. For example, in a 1981 study the British molecular biologist Alec Jeffreys analyzed the DNA sequence of a gene that encodes a protein in the red pigment in the blood of both humans and apes and deduced when the genes began evolving independently in humans and apes. This, I explained, could soon be done for many genes, from many individuals of any species. In this way, scientists would be able to determine how different species were related to one another in the past, as well as when they began their separate histories, with much greater accuracy than was possible from the study of morphology or fossils.

As I explained all this to Rosti, a question gradually arose in my mind. Would this kind of investigation necessarily be restricted to DNA from blood samples or tissues from humans and animals that live today? What about those Egyptian mummies? Could DNA molecules have survived in them—and could they, too, be joined to plasmids and made to replicate in bacteria? Could it be possible to study ancient DNA sequences and thereby clarify how ancient Egyptians were related to one another and to people today? If that could be done, then we could answer questions that no one could answer by the conventional means of Egyptology. For example, how are present-day Egyptians related to Egyptians who lived when the Pharaohs ruled, some 2,000 to 5,000 years ago? Did great political and cultural changes, such as the conquest by Alexander the Great in the fourth century BCE, or by the Arabs in the seventh century AD, result in replacement of a large part of the Egyptian population? Alternatively, were these just military and political events that caused the native population to adopt new languages, new religions, and new ways of life? In essence, were the people who lived in Egypt today the same as those who built the pyramids, or had their ancestors mixed so much with invaders that modern Egyptians were now completely different from their country's ancient population? Such questions were breathtaking. Surely they must have already occurred to someone else.

I went to the university library and searched in journals and books but found no report of any isolation of DNA in ancient materials. No one seemed even to have tried to isolate ancient DNA. Or if they had, they had not succeeded, because if so, surely they would have published their findings. I talked to the more experienced graduate students and postdocs in Pettersson's lab. Given how sensitive DNA is, they argued, why would you expect it to last for thousands of years? The conversations were discouraging, but I didn't give up hope. In my forays into the literature, I had found articles whose authors claimed to have detected proteins in hundred-year-old animal hides in museums—proteins that could still be detected by antibodies. I had also found studies claiming to have detected, under the microscope, the outlines of cells in ancient Egyptian mummies. So *something* did seem to survive, at least sometimes. I decided to do a few experiments.

The first question seemed to be whether DNA could survive for long in tissues after death. I speculated that if the tissue became desiccated, as was the case when a mummy was prepared by the embalmers in ancient Egypt, then DNA might well survive for a long period since the enzymes that degrade DNA need water to be active. This would be the first thing to test. So in the summer of 1981, when not too many people were around in the lab, I went to the supermarket and bought a piece of calf liver. I glued the receipt from the store onto the first page of a new lab book that I would use to record these experiments. I labeled the book with my name but nothing else, since I had decided to keep my experiments as secret as possible. Pettersson might forbid me to pursue them, if they struck him as an unnecessary distraction from the intensely competitive study of the molecular workings of the immune system that I was supposed to be working on. And, in any case, I wanted to keep all this under wraps to spare myself the ridicule of my lab colleagues in the likely event of failure.

To somewhat imitate ancient Egyptian mummification, I decided to artificially mummify the calf liver by sequestering it in an oven in the lab heated to 50°C. The first effect of this was that the secrecy of my project was compromised. By the second day, the repugnant smell elicited considerable comment, and I had to reveal my project before someone found the liver and disposed of it. Fortunately, the smell decreased as the desiccation progressed, and neither the smell nor word of what was putrefying in the lab made it to my professor.

After a few days, the liver had become hard, blackish-brown, and dry—just like an Egyptian mummy. I proceeded to extract DNA from it, with immediate success. The DNA was in small pieces of a few hundred nucleotide pairs instead of the many thousands of nucleotide pairs typical of DNA extracted from fresh tissue, but there was still lots of it. I felt vindicated. It was not totally ridiculous to think that DNA could survive in a dead tissue—at least for some days or weeks. But what about thousands of years? The obvious next step was to try performing the same stunt with an Egyptian mummy. Now my friendship with Rosti came in handy.

Rosti had been primed by my fretting about Egyptology and molecular biology and was happy to abet my attempt to take Egyptology into the molecular age. The small university museum of which he was the curator had some mummies, and he consented to my request to sample them. He was, of course, not about to let me cut them open and remove their livers. But if a mummy was already unwrapped and its limbs had broken off, Rosti allowed me to remove small pieces of skin or muscle tissue from the area where the mummy had already been broken to try my DNA extraction. Three such mummies were available. As soon as I put the scalpel to what had once been the skin and muscles of a person who existed some 3,000 years ago, I realized that the texture of the tissue was different from that of the calf liver I had baked in the oven. The liver had been hard and a bit tough to slice up, whereas the mummies were brittle and their tissues tended to crumble to brown powder when cut. Undeterred, I submitted them to the same extraction procedure I had performed on the liver. The mummy extracts differed from the liver extract in that they were as brown as the mummies themselves, whereas the liver extract had been as clear as water. And when I looked for DNA in the mummy extracts by letting them migrate in a gel in an electric field and staining them with a dye that would fluoresce pink in ultraviolet light if it had bound to DNA, I saw nothing except the brown stuff, which indeed fluoresced in the ultraviolet light, but blue instead of pink, not what was expected if it was DNA. I repeated this process on the two other mummies. Again, there was no DNA; nothing but an undetermined brown substance had ended up in the extracts that I had hoped would contain DNA. My lab colleagues seemed to be right: How could the fragile DNA molecules survive for thousands of years, when even inside a cell it needed constant repair in order not to decompose?

I hid my secret lab book at the bottom of my desk drawer and returned to the virus that tricked the immune system with its clever little protein—but I could not get the mummies out of my mind. How could it be that others had seen what seemed to be remains of cells in some mummies? Perhaps that brown stuff was actually DNA, chemically modified in some way so as to look brown and fluoresce blue in UV light. Perhaps it was naïve to expect DNA to survive in every mummy. Perhaps one needed to analyze many mummies to find the rare good ones. The only way to find out was to convince museum curators to sacrifice pieces of many mummies in the perhaps vain hope that one of them would produce ancient DNA, and I had little idea how to get their permission. It seemed I needed a quick and minimally destructive way to analyze a lot of mummies. My medical education gave me a clue. Very small pieces of tissue, such as those removed with a biopsy needle from a suspected tumor, for example, could be fixed and stained and then studied under a microscope. The level of discernible detail was generally exquisite, allowing a trained pathologist to distinguish normal cells in the lining of the intestine or in a prostate or mammary gland, on the one hand, from cells that had started to change in ways that suggested they were early tumors, on the other. Moreover, there were dyes specific for DNA that could be applied to microscope slides to show whether DNA was present. What I needed to do was to collect small samples from a large number of mummies and analyze them by microscopy and DNA staining. The largest numbers of mummies, obviously, were to be found in the largest museums. But the curators could be expected to be skeptical about letting a slightly overexcited student from Sweden remove even tiny pieces for what seemed a pie-in-the-sky project.

Again, Rosti proved sympathetic; he pointed out that there was one large museum that had huge mummy collections and might be willing to cooperate. It was the Staatliche Museen zu Berlin, a complex of museums in East Berlin, the capital of the German Democratic Republic. Rosti had spent many weeks there working on its ancient Egyptian pottery collection. That Rosti came to East Germany as a professor from Sweden, which at the time was perceived as a country that attempted to find a "third way" between capitalism and communism, probably helped him gain permission to work in the museum. But it was his ability to develop warm friendships across borders that then allowed him to become close friends with several of the curators at the museum. And thus, in the summer of 1983, I found myself on a train that was driven onto a ferry in southern Sweden to arrive the next morning in communist East Germany.

I spent two weeks in Berlin. Every morning I had to pass several security controls to enter the storage facility of the Bode Museum, located on an island in the River Spree near the heart of Berlin. Almost forty years after the war, the museum was still clearly marked by it. On several of the facades, I could see bullet holes in the walls around the windows that had been targeted by machine guns as Berlin fell to the Soviet Army. On the first day, when I was taken to see the prewar Egyptological exhibition, I was handed a hard hat like the ones used by construction workers. It soon became clear why. The exhibition hall had huge holes in the roof from artillery shelling and bombs. Birds were flying in and out, and some were nesting in the pharaonic sarcophagi. Everything that was not of durable material was now sensibly stored elsewhere.

Over the following days, the curator in charge of Egyptian antiquities showed me all his mummies. For a few hours before lunch in his dusty run-down office I removed small snippets of tissues from mummies that were unwrapped and broken. Lunch was a long affair that required exiting through all the security checks to reach a restaurant across the river, where we ate greasy food that needed lubrication with copious amounts of beer and schnapps. Back in the collections, we spent the afternoon over more schnapps, lamenting the fact that the only foreign travel the curator had ever been allowed was visits to Leningrad. It soon became clear that my host dreamed of visiting the capitalist West and that if he got the chance he would probably defect. To provide some perspective on working life in the West I suggested, as diplomatically as I could, that in the west if you drank on the job, you were likely to be fired—an unknown concept in socialism. Such sobering thoughts seemed not to detract from the allure of the opportunities my host imagined to abound in capitalism. In spite of the hours spent on these theoretical discussions, I managed to collect more than thirty mummy samples to take back to Sweden.

At Uppsala, I prepared the samples for microscopy by soaking them in a salt solution to rehydrate them, then mounting them on glass slides and staining them with dyes that permitted visualization of cells. Then I looked for preserved cells in the tissues. I did this work on weekends and late at night, so as not to let it be widely known what I was doing. As I peered through the microscope, the appearance of the ancient tissues depressed me. In muscle sections, I could barely discern the fibers, let alone any traces of cell nuclei where DNA might be preserved. I was almost despairing, until one night I looked at a section of cartilage from a mummified outer ear. In cartilage, as in bone, cells live in small holes, or lacunae, inside the compact, hard tissue. When I looked at the cartilage, I saw what appeared to be the remains of cells inside the lacunae. Excited, I stained the section for DNA. My hands were trembling as I put the slide under the microscope. Indeed, there was staining within the cellular remains in the cartilage (see Figure 2.1). There seemed to be DNA preserved inside!

With renewed energy, I went on to process all of the remaining samples from Berlin. A few looked promising. In particular, the skin from the left leg of the mummy of a child showed what were clearly cell nuclei. When I stained a section of the skin for DNA, the cell nuclei lit up. Since this DNA was in the cell nuclei, where the cellular DNA is stored, it could not possibly be from bacteria or fungi because such DNA would appear at random in the tissue where the bacteria or fungi were growing. This was unambiguous evidence that DNA from the child herself was preserved. I took many photos through the microscope.

I found three mummies with staining of the cell nuclei showing the presence of DNA. The child seemed to have the largest number of

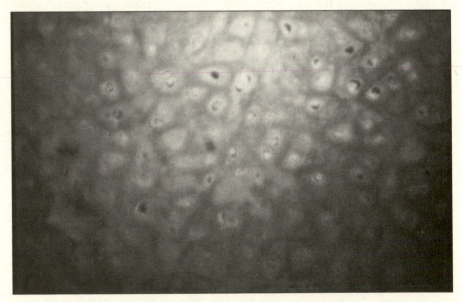

FIGURE 2.1. Microscopic picture of cartilage tissue from an Egyptian mummy from Berlin. In some lacunae, cell remains light up suggesting that DNA may be preserved. Photo: S. Pääbo, Uppsala University.

well-preserved cells. But now doubt started to gnaw at me. How could I be sure that this mummy was really old? Modern corpses were sometimes falsified to look like ancient Egyptian mummies so that the perpetrators might earn a few dollars from tourists and collectors. Some of these mummies might later be donated to museums. The staff of the museum in Berlin had been unable to give me any records of the provenance of this particular mummy, perhaps because the relevant parts of the catalog had been destroyed in the war. The question of its age could be resolved only through carbon dating. Fortunately, Göran Possnert, an expert on carbon dating, worked at Uppsala University. He used an accelerator to determine the ages of tiny samples of ancient remains by measuring the ratios of carbon isotopes present. I asked him how much it would cost to date my mummy, worrying that I would not be able to afford it on my meager student stipend. He took pity on me and offered to date it for free, considerately not even mentioning the price, doubtless because it would have been well out of my range. I delivered a small piece of the mummy to Göran and waited to hear the results. For me, this exemplified one of the most frustrating situations in science, when your work depends crucially on the work of someone else and you can do nothing to expedite it—just wait for a phone call that seems to never come. But finally, a few weeks later, I got the call I had been waiting for. The news was good. The mummy was 2,400 years old; it dated from about the time of the Alexandrian conquest of Egypt. I drew a sigh of relief. First I went out and bought a big box of chocolates, which I delivered to Göran. Then I started to think about publishing my findings.

During my time in East Germany, I had developed some understanding of the sensitivities of people living under socialism. In particular, I knew that the museum curator and other museum officials who hosted my visit would be very disappointed with just a perfunctory expression of gratitude at the end of my paper. I wanted to do the right thing, so after talking to Rosti and conferring with Stephan Grunert, a young and ambitious East German Egyptologist whom I had befriended in Berlin, I decided to publish my first paper on mummy DNA in an East German scientific journal. Struggling with my high school German, I wrote up my findings, including photographs of the mummy itself and of the tissue stained for DNA. In the meantime, I had also extracted DNA from the mummy. This time, the extracts contained DNA that I could demonstrate in a gel, and I included a picture of such an experiment in the paper. Most of the DNA was degraded, but a small fraction of it was several thousand nucleotides long, similar in length to the DNA one could extract from fresh blood samples.

This, I wrote, seemed to indicate that some DNA molecules from ancient tissues might well be large enough to allow the study of individual genes. I speculated wildly about what might be possible if DNA from ancient Egyptian mummies could be systematically studied. The paper ended on a hopeful note: "Work over the next few years will show if these expectations will be fulfilled." I sent the manuscript to Stephan in Berlin. He fixed up my German, and in 1984 the article appeared in *Das Altertum,* a journal published by the East German Academy of Sciences.[1] And nothing happened. Not a single person wrote to me about it, much less asked for a reprint. I was excited, but no one else seemed to be.

Having realized that the world at large did not make a habit of reading East German publications, I had written up similar results from the fragment of the mummified head of a man and, in October of the same year, had sent them to a Western journal that seemed appropriate—the *Journal of Archaeological Science.* But here the frustration turned out to be the unbelievable slowness of the journal, even compared with the delay my manuscript had experienced in East Germany, where it needed to be fixed up linguistically by Stephan and then presumably scrutinized by the political censors. This was, I felt, a reflection of the glacial speed with which the disciplines concerned with ancient things were moving. The *Journal of Archaeological Science* finally published my paper at the end of 1985[2]—by which time the results it described had been largely overtaken by events.

The next step—now that I had some mummy DNA—was obvious. I needed to clone it in bacteria. So I treated it with enzymes that make the ends of the DNA amenable to being joined to other pieces of DNA, mixed it with a bacterial plasmid, and added an enzyme that joins DNA fragments together. If successful, this would create hybrid molecules in which pieces of DNA from the mummy were joined to the plasmid DNA. When these plasmids were introduced into bacteria, they would not only allow the hybrid molecules to replicate to high copy numbers in bacterial cells but would also make the bacteria resistant to an antibiotic I would add to my culture medium, so that the bacteria would survive only if they contained a functioning plasmid. When seeded on growth plates containing the antibiotic, colonies of bacteria would appear if the experiment was successful. Each such colony would derive from a single bacterium that now carried one particular piece of mummy DNA. To check on my experiment, I did controls—an essential thing in any laboratory experiment. For example, I repeated the exact process in parallel but added no mummy DNA to the plasmid, and also repeated the process but added modern human DNA. After making the bacteria take up the DNA solutions from these experiments, I plated them on agar plates containing the antibiotic and put them in an incubator at 37°C overnight. The next morning I opened the incubator and, with anticipation, inhaled the puff of moist air smelling of rich culture media. The plate with the modern DNA yielded thousands of colonies, so many that it was almost totally covered with bacteria. This showed that my plasmid had worked: the bacteria survived because they had taken up the plasmid. The plate where no DNA had been added to the plasmid yielded hardly any colonies, indicating that I did not have DNA from some unknown source in my experiment. The experiment itself, where I had added the DNA from the Berlin mummy, yielded several hundreds of colonies. I was ecstatic. I had apparently replicated 2,400-year-old DNA! But could it have come from bacteria in the child's tissues, rather than from the child herself? How could I show that at least some of the DNA I had cloned in the bacteria was human?

I needed to determine the DNA sequence from some of the DNA in order to show that it was human rather than bacterial. But if I merely sequenced random clones, they would be likely to contain DNA sequences that could have come either from the human genome—which in 1984 was not yet decoded, except for some tiny parts that had been sequenced with great effort—or from some microorganism whose DNA sequences were even less likely to be known. So instead of sequencing random clones, I needed to identify some clone of interest. The answer lay in a technique whereby one could identify clones that carried DNA similar in sequence to something one wanted to find. This technique involved transferring some of the bacteria from each of hundreds of colonies to cellulose filters, where the bacteria were broken open and their DNA were bound to the filter. I then used a radioactively labeled piece of DNA, a "probe," that was single-stranded and then hybridized to complementary sequences from the single-stranded DNA on the filters. I chose to use a piece of DNA that contains a repeated DNA element—the so-called *Alu* element—of about 300 nucleotides that occurs almost a million times in the human genome and in no organisms besides humans, apes, and monkeys. In fact, these *Alu* elements are so numerous that more than 10 percent of the human genome is made up of them. If I could find an *Alu* element among my clones, it would show that at least some of the DNA I had extracted from the mummy came from a human being.

I got a piece of a gene I'd studied in the lab that contained an *Alu* element, incorporated radioactivity in it, and hybridized it to my filters. Several of the clones took up the radioactivity, as one would expect if some of the DNA was human. I picked the most strongly hybridizing clone. It contained a piece of DNA consisting of about 3,400 nucleotides. With the help of Dan Larhammar, a graduate student who was the master of DNA sequencing in our group, I sequenced a part of the clone. It did indeed contain an *Alu* element. I was very happy. There was human DNA among my clones, and it could be cloned in bacteria.

As I was grappling with my sequencing gels in November 1984, a paper appeared in *Nature* that was of great relevance for me. Russell Higuchi, who worked at UC Berkeley with Allan Wilson, the primary architect of the out-of-Africa theory of modern human origins and one of the most famous evolutionary biologists of the time, had extracted and cloned DNA from the 100-year-old skin of a quagga, an extinct subspecies of zebra that had existed in southern Africa until about a hundred years ago. Russell Higuchi had isolated two fragments of mitochondrial DNA and shown that the quagga was, as expected, more closely related to zebras than to horses. This work inspired me greatly. If Allan Wilson was studying ancient DNA, and if *Nature* considered an article about 120-year-old DNA interesting enough to publish, then surely what I was doing was neither crazy nor uninteresting.

For the first time, I sat down to write a paper of my own that I believed many people in the world would be interested in. Inspired by Allan Wilson's example, I wrote it for *Nature*. I described what I had done with the mummy from Berlin. One of my first references was to the paper that had appeared in the East German journal. However, before I sent the manuscript off to London, where *Nature* had its office, there was something I needed to do. I needed to talk to my thesis adviser, Per Pettersson, and show him the manuscript, now ready to submit. With some trepidation, I entered his office and told him what I had done. I asked if he might perhaps want to be a co-author with me on the paper, in his capacity as my adviser. Obviously, I had underestimated the man. Rather than scolding me for what could have been seen as misappropriation of research funds and valuable time, he seemed amused. He promised to read the manuscript and said that, no, obviously he should not be the co-author of work that he hadn't even been aware of.

A few weeks later, I received a letter from *Nature,* with a promise from the editor to publish my manuscript if I could respond to some minor comments from reviewers. Shortly thereafter, the proofs arrived. At that point, I thought about how to approach Allan Wilson—a demigod, in my view—to ask if I might work with him at Berkeley after my PhD defense. Not knowing exactly how to broach this topic, I mailed him a copy of the proofs without any comment whatsoever, thinking he might appreciate seeing the paper before it appeared. I thought that I would then later write to him about job opportunities in his laboratory. *Nature* progressed rapidly toward publication and even solicited a cover illustration of a mummy with DNA sequences artfully wrapped around it. Even more rapidly, I received a response from Allan Wilson, who addressed me as "Professor Pääbo"—this was before both the Internet and Google, so there was no obvious way for him to find out who I was. The rest of his letter was even more amazing. He asked if he could spend his upcoming sabbatical year in "my" laboratory! This was a hilarious misunderstanding, resulting from my insecurity about knowing what to write to him. I joked with my lab mates that I would have Allan Wilson, perhaps the most famous molecular evolutionist of the time, wash gel plates for me for a year. Then I settled down to write him back— explaining that I was not a professor, not even a PhD, and certainly did not have a lab where he could spend his sabbatical. Rather, I wondered if there might be a chance for me to spend my postdoc in his Berkeley lab.

Chapter 3
Amplifying the Past

Allan Wilson wrote me a gracious reply, inviting me to work in his group as a postdoctoral fellow. This would prove to be a turning point in my career. Once I had earned my PhD, I had three choices: finishing my medical studies at the hospital (a boring prospect after the excitement I had just experienced); following up on my successful PhD work on viruses and immune defense at some world-class lab; or accepting Allan's offer to spend my postdoc trying to retrieve ancient genes. Most of my peers and the professors with whom I discussed these choices suggested the second alternative, arguing that my interest in mummy DNA was a quaint hobby but ultimately a distraction from the serious work on which a solid future in research could be built. I, of course, was tempted by the third option but still felt hesitant, wondering whether mainstream research in virology, with "molecular archaeology" as a hobby, was not the more realistic course. What changed it all was the 1986 Cold Spring Harbor Symposium.

Cold Spring Harbor Laboratory, on Long Island, New York, is the Mecca of molecular genetics. The laboratory organizes many well-respected meetings, in particular a yearly Symposium on Quantitative Biology. Thanks to my paper in *Nature*,[1] I was invited to the 1986 symposium, where I presented, for the first time, a lecture on my mummy work. As if this were not already exciting enough, in the audience were many people I knew only from the literature, including Allan Wilson himself and Kary Mullis, who in the same session described the polymerase chain reaction. The PCR was a real technical breakthrough, since it did away with most of the cumbersome cloning of DNA in bacteria, and it was immediately obvious to me that it might be used to study ancient DNA. In principle, the PCR would enable me to target and multiply DNA segments of interest even if just a few survived. In fact, referring to my presentation, Kary ended his talk by noting that the PCR would be ideally suited for studying mummies! I could hardly wait to get back to the lab and try it out.

The meeting was electrifying in another way as well: it was the first time that a coordinated and publicly funded effort to sequence the entire human genome was on the agenda. Although the meeting made me feel much like the novice I was, I was elated to be present as the big guys discussed the millions of dollars, the thousands of machines, and the new technologies needed for this endeavor. In lively debates, some well-known scientists denounced the proposed project as technically impossible, unlikely to yield interesting results, and likely to divert valuable funding from more worthwhile research by small groups led by single investigators. To me, it was all very exciting; I wanted to be part of the genomic adventure.

Unlike most of the testosterone-fueled, high-powered scientists dominating the meeting, Allan Wilson was low-key and soft-spoken, the personification of what I imagined a Berkeley don to be. A long-haired New Zealander with a warm gaze, he made me feel comfortable and encouraged me to follow my inclinations and do what seemed most promising to me. The meeting with him helped me overcome my indecision and I told him I wanted to come to Berkeley.

There was a hitch, though. Unable to come to "my" laboratory for his sabbatical, Allan had decided to spend the year at two labs in England and Scotland, which meant that I would have to find something else to do in the meantime. As a part of my PhD project, I had worked for a few weeks in the Zurich laboratory of Walter Schaffner, a famous molecular biologist who had discovered "enhancers," crucial DNA elements that help drive the expression of genes. Walter, always full of unabashed enthusiasm for unorthodox ideas and projects, now invited me to spend the year in his lab working on ancient DNA. He was particularly interested in the thylacine, an extinct wolf-like marsupial from Australia. Could I not clone DNA from museum specimens of this creature? I agreed and moved to Zurich as soon as I had passed my PhD defense in Uppsala.

In the meantime, I had hoped that the attention generated by my *Nature* paper would allow me to obtain more mummy samples from East Germany, so that I could generate more clones and find interesting genes instead of the mundane *Alu* repeats. So when Rosti went to Berlin some months after the *Nature* publication to arrange for me to sample the mummies again, I expected clear sailing. Instead, he returned with disturbing news. None of his friends at the museum had had time to see him; in fact, they all seemed to avoid him. Eventually he had been able to corner one of them as the fellow was leaving the museum. It seems that after the publication of my

Nature paper, the Stasi, the feared East German secret police, had appeared at the museum and interviewed each staff member in turn in a small room, asking them what they had been doing with me and Rosti. That I had published my first results in East Germany and had prominently referred to that publication in the *Nature* paper—none of this impressed the Stasi. Instead, they impressed upon the museum employees that, as they put it, Uppsala University is a well-known antisocialist propaganda center. No matter how ridiculous this characterization of the oldest university in Sweden was, no East German citizens in their right mind would of course have anything to do with us after being told this by the Stasi.

I was depressed by the futility of dealing with a totalitarian system. Having entertained visions that our two competing political systems might grow closer, perhaps catalyzed by scientific contacts, I had hoped that I might contribute to the process in some small way. Little did I know the role that East Germany would play in my life, but at that point neither samples nor cooperation seemed to be in the cards.

In Zurich, I set about extracting DNA both from the small mummy samples I had left in my possession and from specimens of the marsupial wolf. Despite my enthusiasm for the PCR, getting it to work following Kary Mullis's protocol was no picnic. It involved heating the DNA in a 98°C water bath to separate the strands, then cooling it in a 55°C water bath to let the synthetic primers attach to their targets, then adding the heat-sensitive enzyme and incubating the mix in a 37°C water bath to try and coax it to make the new strands. For each experiment, this tedious cycle of manipulations needed to be done at least thirty times. I spent hours on end in front of steaming water baths wasting many test tubes of expensive enzyme in my attempts to amplify pieces of DNA. Sometimes I was able to generate a weak product from modern DNA, but I had no luck with the badly degraded DNA from the thylacine and the mummy samples. I did have some success in showing, by electron microscopy, that much of the mummy and thylacine DNA was in short pieces. Some DNA molecules had even become linked to each other by chemical reactions, a feature that was sure to make them intractable to multiplication either in bacteria or by PCR in the test tube. This was not surprising, given some findings I had made in 1985, when I had visited Tomas Lindahl's lab in Hertfordshire outside London for a few weeks. Tomas is originally of Swedish descent and one of the world's experts on chemical damage to DNA and the systems that organisms have evolved to repair it. In his lab I had shown that

there was evidence for several forms of damage in the DNA I had extracted
from the old tissues. These results as well as my new Zurich findings constituted solid descriptive science, but they did not take me closer to my goal
of reading DNA sequences from long-extinct creatures. Months passed in
front of the water baths—as well as on the Alpine ski slopes—but no breakthroughs transpired, so it was with a distinct sense of relief in the spring
of 1987 that I left Zurich for Berkeley, where Allan Wilson was back in
residence.

Upon arriving in the Biochemistry Department at UC Berkeley, I soon realized that I was in the right place at the right time. Kary Mullis had been
a graduate student there before he moved to the Cetus Corporation, down
by the Bay, where he invented the PCR. Several of Allan's previous graduate students and postdocs worked at Cetus. The result was that while I had
been alone in my struggle to get the PCR to work in Zurich, in Berkeley
many people worked on it, and as a result many improvements were made.
At Cetus, they had cloned and expressed a version of DNA polymerase,
the enzyme used in the PCR to make new DNA strands, from a bacterium
that grows at high temperatures. Since this enzyme could survive high
temperatures, there was no need to open the test tubes and add enzyme
during each PCR cycle. This meant that now the whole process could be
automated; indeed, one postdoc in the lab had already built a contraption
in which a small water bath was fed water from three bigger water baths in
cycles controlled by a computer. This allowed the PCR to be done automat-
ically. After months in front of the water baths in Zurich, this was progress
I could appreciate. I could start a PCR and leave for home in the evening (a
practice my colleagues and I had to abandon after a major flood in the lab
when a valve failed to close as expected). Our innovative but unreliable lab
machinery was soon replaced by the first PCR machine produced by Cetus.
Consisting of a metal block with holes for the test tubes, it would heat and
cool our samples however we pleased, for as many cycles as we wanted, all
of this computer-controlled. I remember the awe we all felt when it was
wheeled in. I threw myself at this machine, booking it for as many runs as
my lab mates would tolerate.

The extinct South African zebra, the quagga, from which Russell Higuchi
had cloned two pieces of mtDNA, offered itself as a first step. Russ had left

Allan's lab for Cetus, but some of his quagga samples were still there. I extracted DNA from a piece of quagga skin, had primers synthesized for the same mitochondrial sequences he had cloned, and started a PCR in the new machine. It worked! I amplified beautiful pieces of quagga DNA, and when I sequenced them, they were very similar to what Russell had determined by cloning in bacteria. The big advance was that I could do it again and again. Bacterial cloning was so inefficient that replication of the findings would have been next to impossible, because the process was unlikely to produce the same stretch of DNA. The quagga sequences I retrieved were very similar to the sequences cloned in bacteria but they differed in two places from Russell's sequences, probably because of molecular damage that had induced errors when the bacteria took up and replicated his samples. With the PCR, I could now retry the same sequence multiple times to ensure that it could be exactly replicated. This was what science was supposed to be about: reproducibility of results!

I published the quagga data in *Nature,* with Allan as my co-author.[2] Clearly it was now possible to study ancient DNA in a systematic and controlled way. I felt sure that extinct animals, Vikings, Romans, Pharaohs, Neanderthals, and other human ancestors would now soon be subject to the powerful methods of molecular biology, although proving that would take some time. (After all, I had to compete with my lab mates for use of our PCR machine.) One interest of Allan's was human origins. Not much earlier, with Mark Stoneking and Rebecca Cann, he had published a controversial paper in *Nature* comparing mitochondrial DNAs from people from all over the world by means of cumbersome analyses using enzymes that cut the DNA at various places of known sequence—indicating that the mtDNAs could be traced back to a single common ancestor, who lived in Africa some 100,000 to 200,000 years ago.[3] Now this work could be extended by studying DNA sequences from many more individuals. A young graduate student, Linda Vigilant, who arrived at the lab on a motorcycle every morning, was doing this work. I was peripherally aware of her boyish charm but saw her mostly as a competitor for time on the coveted PCR machine. Little did I know that at a later time and in another country, we would be married and have children together.

So far, the reconstruction of human evolution from genetic data had been limited to studying differences in DNA sequences in living individuals and inferring how past migrations had resulted in the differences. These inferences were based on models that reflected ideas about how DNA sequences accumulate nucleotide changes and how variants are transmitted from generation to generation in populations, but the models were by necessity great oversimplifications of what could have gone on in the past. They assumed, for example, that within a population every individual had an equal chance to produce children with every other individual of the opposite sex. They also assumed each generation to be a discrete entity with no intergenerational sex and no difference in survival based on the DNA sequences under study. Sometimes I felt that this amounted to little more than making up stories about the past, and very clearly all of it was indirect. To go back in time and actually see what genetic variants had existed in the past would be "catching evolution red-handed," as I liked to say, by studying DNA sequences from many individuals in the past, adding direct historical observations to the work that Linda was doing on people living today.

These were ambitious ideas, so I decided to try them out over time periods more modest than thousands of years. The Museum of Vertebrate Zoology at UC Berkeley housed enormous collections of small mammals assembled by naturalists working in the American West over the past hundred years. With Francis Villablanca, a graduate student from the museum, and Kelley Thomas, a postdoc in Allan's lab, I set out to study populations of kangaroo rats, small rodents named for their tendency to jump around on their inordinately large hind legs (see Figure 3.1). They are abundant in the Mojave Desert on the border between California, Nevada, Utah, and Arizona, where they are the favorite food of rattlesnakes. I extracted and sequenced mtDNA from the skins of several in the museum that had been collected at three different places in 1911, 1917, and 1937. Francis, Kelley, and I then got copies of the zoologists' field notes and maps and took off for a series of trips to the Mojave to set traps at the same locations. We drove into the desert following the old field maps and identified the places where our zoologist predecessors had been forty to seventy years earlier. As the sun was setting, we set traps among sage brush and Joshua trees. Sleeping under the stars on clear and calm desert nights, interrupted only occasionally by the sound of our rodent traps snapping shut, was a pleasurable change from my urban, work-filled, everyday life.

Back in the lab, we extracted and sequenced mtDNA from the rodents we had collected and compared them to the sequences from animals that had lived some forty to seventy generations before. We found that these variants had not changed markedly over time, and while this observation was not entirely unexpected, it was still satisfying in that this

FIGURE 3.1. A hundred-year-old kangaroo rat and a present-day one from the Museum of Vertebrate Zoology at UC Berkeley. Photo: UC Berkeley.

was the first-ever peek back in time at the genes of the ancestor populations of animals living today. We published our findings in the *Journal of Molecular Evolution*[4] and were pleased to find a glowing comment about our work in *Nature*[5] by up-and-coming evolutionary biologist Jared Diamond, who said that the new techniques made possible by the PCR meant that "old specimens constitute a vast, irreplaceable source of material for directly determining historical changes in gene frequencies, which are among the most important data in evolutionary biology." He also said that "this demonstration project will make life harder for those who are too narrow-minded to understand the scientific value of museum specimens."

However, to me, human evolutionary history was the Holy Grail, and I wondered whether the PCR could open a window into our own past. In Uppsala, I had gotten a sample from some gruesome yet amazing discoveries made in Florida sinkholes. In these water-filled alkaline deposits, ancient Native American skeletons were found; inside the crania, the brains, although slightly shrunken, were preserved in surprising detail. Using old-fashioned techniques, I had shown that the sample contained preserved human DNA, and I presented these results at Cold Spring Harbor, along with my mummies. Through Allan, I now acquired a sample from a similar find in Florida of 7,000-year-old brains. I extracted DNA and retrieved short fragments that appeared to be an unusual mtDNA sequence that existed in Asia but that had not previously been seen in Native Americans. Although I found the sequences twice in independent experiments, I had realized by now that contamination with modern DNA was a very common problem, particularly when ancient human remains were being studied. So I cautioned in the paper that "indisputable proof that the amplified human sequences reported here are of ancient origin must await more extended work."[6]

Nevertheless, this research seemed promising; perhaps I needed to learn more about human population genetics. When Ryk Ward, a theoretical population geneticist from New Zealand who worked in Salt Lake City, contacted Allan's lab to learn about the PCR, I volunteered to work with him. This resulted in a monthly commute to Utah, where I taught people in Ryk's lab how to do PCRs. Ryk, an excellent population geneticist, was also pleasantly eccentric, given to wearing shorts and knee socks even in cold weather and taking on projects and various administrative tasks without finishing them. This latter habit did not endear him to his university, but on the plus side he loved to discuss science and had an almost infinite patience for explaining complicated algorithms to people like me who sadly lacked formal mathematical training. Together we studied mtDNA variation in the Nuu-Chah-Nulth, a small First Nations group on Vancouver Island with whom Ryk had worked for many years. Amazingly, we found that the few thousand individuals in this group contained almost half the mtDNA variation that exists among native people throughout the North American continent. This finding suggested to me that the common belief that such tribal groups in the past were genetically homogeneous was a myth, and that instead humans may always have lived in groups that contained substantial amounts of genetic diversity.

Back in Berkeley, it seemed that almost everything we tried worked. When Richard Thomas, a Canadian postdoc, came to learn PCR in the lab and needed a project, I suggested he take a turn working on *Thylacinus cynocephalus,* the marsupial wolf that had frustrated me during my sojourn in Zurich. The thylacine, native to Australia, Tasmania, and New Guinea, looked very much like a wolf but was a marsupial, like kangaroos and several other Australian animals. It was therefore a textbook example of convergent evolution, the process whereby unrelated animals in similar environments and subject to similar pressures often evolve similar forms and behaviors. By sequencing small pieces of mtDNA from the marsupial wolf, we showed that it was closely related to other carnivorous marsupials in the region, such as the Tasmanian devil, but distant from South American marsupials, although some extinct marsupials there had been very wolf-like. This meant that wolf-like animals evolved not only twice but three times, once among placental mammals and twice among marsupials. Thus evolution was, in a sense, repeatable—an observation that had already been made, and would be made again, in studies of other groups of organisms. We wrote this up for *Nature,* and Allan graciously allowed me to be last author, the place occupied by the scientist who has led the work.[7] This was a first for me, and I knew that my situation in science was beginning to change. Until now, I had been someone who did the work at the lab bench, producing results by doing experiments myself the whole day and often much of the night; even when the ideas were my own I was often helped and inspired by discussions with a supervisor. Now I realized that this was beginning to change. I was not doing all experiments myself anymore. Gradually, I would have to be the one to lead and inspire others. While this prospect seemed daunting when I thought about it in the abstract, I nevertheless found that doing so often came naturally to me.

Whereas I worked with others on many applications of the PCR to ancient DNA, I concentrated my own efforts on understanding the technical intricacies of ancient DNA retrieval. I summarized the knowledge accumulated during my work in Uppsala, Zurich, London, and Berkeley in a paper in the *Proceedings of the National Academy of Sciences,* showing that DNA in ancient remains was generally short in length, contained many chemical modifications, and sometimes exhibited cross-links between molecules.[8] The degraded state of the DNA had several implications for work done with the PCR. Its main consequence was the unfeasibility of using the PCR to retrieve long pieces of ancient DNA. Anything above 100 or 200 nucleotides was generally impossible. I also found that when there were few or even no molecules long enough for the DNA polymerase to operate continuously from one primer to the other, the polymerase would sometimes stitch shorter pieces of DNA together, producing Frankenstein's monster–like combinations that did not exist in the original genome of the ancient organism. Formation of such hybrid molecules through this process, which I called "jumping PCR," is an important technical complication that can confuse results when amplifying ancient DNA, and I described it in two papers, but I totally overlooked its broader implications. As it happens, a few years later the same basic stitching process was used by a more practically oriented scientist, Karl Stetter, to combine pieces of different genes to generate new "mosaic" genes that made proteins with new properties. This idea—which, being totally focused on my forays into the past, I had utterly failed to consider—formed the basis for a whole new branch of the biotech industry.

While many things were working well in Allan's lab, the limitations of the new techniques and of DNA preservation also began to be discernible to me. First, not all ancient remains contained DNA that could be retrieved and studied, even by the PCR. In fact, apart from museum specimens that had been prepared rapidly after an animal's death, few old samples yielded DNA that could be amplified. Second, in old samples that did yield DNA, its degraded state meant that one could generally amplify only pieces that were 100 or 200 nucleotides long. Third, it was often next to impossible to amplify nuclear DNA from old specimens. My dream at Uppsala of finding long pieces of nuclear ancient DNA seemed to be just that—a dream.

My life in the Bay Area was intense and satisfying, outside as well as inside the lab. I had always been attracted to men as well as women and had been active in the gay-rights movement in Sweden. In the Bay Area the AIDS epidemic was growing exponentially and took the lives of thousands of young men. Feeling I had to do something to help, I had joined the AIDS Project of the East Bay as a volunteer. There I encountered two of the most beautiful aspects of American society: self-organization and volunteerism, habits often lacking in Europe. Yet, in spite of this welcoming atmosphere and the scientific opportunities I encountered in the United States, I wanted to return to Europe eventually. It was a girlfriend who was to have a decisive influence on the route the rest of my life would take. Barbara Wild, a German graduate student in genetics, was visiting Berkeley, and I was introduced to her by Walter Schaffner, who had arranged her stay there. She was energetic, beautiful, and smart. We had a short but intense affair that continued even after she returned to her native Munich. I took every opportunity to visit Europe; on one occasion we met for an almost ridiculously romantic weekend in Venice. Much of my emotional life since my teenage years had been centered on infatuations with heterosexual men, many of whom ended up as little more than friends. It was an exhilarating experience to walk around in Venice with Barbara and behave publicly in a way I had never dared with my erstwhile boyfriends.

To give my frequent trips to Munich a veneer of science, I paid several visits to the Genetics Department of the Ludwig-Maximilians-Universität, where Barbara was a graduate student. Once, I even gave a seminar there on my experiments with ancient DNA. After my seminar the molecular biologist Herbert Jäckle asked me if I would be interested in an assistant professorship that would become available there in a couple of months. I said yes, seeing an opportunity to spend time on a more permanent basis with Barbara. But on a subsequent visit to Munich I realized that she had become involved with another scientist—one who studied fruit flies, as she did. Indeed, he would later become her husband. I flew back to Berkeley and did my best to forget about Barbara and Munich.

Six months later, I started to apply for jobs in earnest. I visited Cambridge University, where I was offered a lectureship; I visited Uppsala, where I was offered a position as a research assistant. Then, late one night, Germany caught up with me again, in the form of Charles David, the American-born dean of the biology faculty in Munich, who called me in Berkeley. Would I consider coming to Munich if the university offered me a full professorship instead of an assistant one?

This would amount to a tremendous advance in my career. Normally one would expect to be an assistant professor for a number of years before becoming a full professor. The full professorship was not just a title but came with resources such as a big laboratory, personnel, and funds. Yet I hesitated. I knew little of Germany, except for its reputation as home to one of the two worst political ideologies of the century. I had no idea whether I would fit in, or if my bisexuality would create problems. Ultimately, Charlie and Herbert together convinced me that Munich was a place where one could both live well and do science, so I decided to give it a try. My plan was to take the opportunity offered by the Munich professorship and do good work there for a few years so that I could then move back to Sweden. I accepted their offer and arrived in Munich with two big suitcases early one morning in January 1990, prepared to begin my independent scientific life in a world new to me and more than a little frightening.

Chapter 4
Dinosaurs in the Lab

Setting up a laboratory is an intimidating experience, especially the first time you do it, even more so when you're doing it in an unfamiliar environment. In my case, the environment was new in more than one sense. First of all, it was loaded with German history. The building where I was to work, the university's Institute of Zoology, had been built and donated to the university by the Rockefeller Foundation during the Great Depression of the 1930s. During the war, it was bombed by the Americans, and it was rebuilt by the foundation after the war, so it epitomized the complicated and multifaceted relationship between Germany and the United States, a pendulum that has swung back and forth between the extremes of war and alliance. The institute was situated between the railway station and the complex of buildings erected by Hitler to house the Nazi Party headquarters. It was rumored that below the basement was a tunnel once used by the führer and his associates to move from the station to the headquarters. True or not, the rumor symbolized my fears of latent, subsurface German fascism.

Another novel aspect of the environment was that my appointment was to the Zoology Institute. I had never studied zoology, or even biology at the university level—just medicine, since in Sweden you can enter medical school directly after high school. This lapse became all too apparent almost as soon as I arrived, when an older professor asked whether I could perhaps teach the course on insect taxonomy in the upcoming semester. I was still jet-lagged, and preoccupied with other concerns, so without much thought I expressed my surprise that a zoology institute would deal with insects, as insects were hardly animals. In my mind, "animals" were things with paws, fur, and preferably floppy ears. The professor stared at me in disbelief and left without a word. I was immediately ashamed of having made such a complete fool of myself in the first week of my new job. But the good news was that no one ever again suggested that I teach any form of taxonomy or entomology at the institute.

As I was settling in, I learned that my predecessor at the institute had died unexpectedly of food poisoning. It was obviously not going to be easy to win the loyalty of all his colleagues, some of whom viewed me as an inexperienced and eccentric foreigner—a usurper of sorts. This was made clear in an unnerving encounter with Hansjochem Autrum, an emeritus professor and my predecessor's mentor. Professor Autrum had been an influential figure in German zoology; when I arrived in Munich he still edited *Naturwissenschaften,* a somewhat influential German biology journal, and had an office on the same floor as my lab. During my first days in Munich, when I passed him on the staircase I greeted him cordially, but I got no response. One of my technicians reported that afterward he was heard to loudly complain that many good young German scientists could not find jobs and what did the department do but hire "international trash" (*internationaler Schrott*). I decided to ignore him from then on. Many years later, after his death, I became a member of a prestigious German order to which he had belonged, and I read his obituary in its proceedings. The author pointed out that before 1945 Professor Autrum had been a member not only of the Nazi Party but also of the Stormtroopers (SA) and had taught National Socialist ideology courses at a university in Berlin. Although I generally have a somewhat exaggerated desire to be liked by everybody, I felt retrospectively justified in having failed to befriend him.

Fortunately, Professor Autrum was an exception at the institute. Equally fortunately, he represented a generation that was on its way out in Germany. Gradually, by being frank about my ignorance not only of taxonomy but of most things zoological and administrative, I succeeded in bringing around even the older technicians in my group, and soon they wanted to help me build something that would be new and exciting. Charlie and Herbert, for their part, were extremely supportive. When the required laboratory renovations became more expensive than expected, the university came up with additional money. Slowly but surely, the equipment I needed was assembled, and all was put in order. Even more important, some students expressed interest in working with me.

Scientifically, I felt we needed to get systematic about establishing reliable procedures for amplifying ancient DNA. In Berkeley, I had begun to realize that contamination of these kinds of experiments with modern DNA was a serious problem, especially when the PCR was used. With the new PCR machines and the heat-resistant DNA polymerase, the process was sensitive enough that under favorable circumstances a handful of DNA molecules, or perhaps even a single molecule, could start the reaction. That sounds wonderful, but could lead to trouble. If, for example, a museum specimen contained no surviving ancient DNA but a few DNA fragments from some museum curator, we could unwittingly wind up studying the curator's DNA instead of the DNA of an ancient Egyptian priest. Extinct animals, of course, presented much less of an opportunity to mislead ourselves; in fact, it was in the course of doing such work that I first realized the huge potential for contamination, since sometimes when I tried to amplify mtDNA from animal remains, I would get human mtDNA sequences instead. In 1989, shortly before I left Berkeley for Munich, I had published a paper with Allan Wilson and Russell Higuchi, whose quagga work I replicated, in which we introduced what we called criteria of authenticity; these were procedures we thought had to be carried out before a DNA sequence retrieved by the PCR could be confirmed as truly old.[1] We recommended that a "blank extract"—that is, an extract with no ancient tissue but containing all other reagents to be used—be processed in parallel every time extractions from old specimens were performed. This allowed us to detect DNA that might lurk in the reagents themselves, which came to the lab from various suppliers. In addition, extractions and PCRs needed to be repeated several times, to ensure that a DNA sequence could be replicated at least twice. And finally, I had realized that hardly any fragments of ancient DNA were longer than 150 nucleotides. In short, I had concluded that many experiments purporting to have isolated ancient DNA that had been done up to that point, and especially before the PCR became available, were hopelessly naïve.

In hindsight, I now realized that the mummy sequence I had published in 1985 was suspiciously long given that my subsequent work had shown that ancient DNA was almost always degraded to small fragments. One of two factors could explain why the sequence I found, as another group demonstrated, came from a transplantation antigen gene[2] (precisely the type of genes we had studied in our lab back in Uppsala): either because I had identified the sequence with a probe for such genes or because a piece of DNA from the lab had contaminated my experiments. Given the length of the sequence, contamination seemed much more likely. I consoled myself with the thought that this is how science progresses: older experiments are overtaken by new and better ones. And I was happy to be the one to improve on my own work. With time, there also came help from outside the field. In 1993, Tomas Lindahl published a short comment in *Nature* in which he suggested that criteria much like the ones we championed in 1989[3] were necessary for the ancient DNA field.[4] It was a great help to have a respected scientist from outside the field point this out—especially given my concern that the ancient DNA field tends to attract people without a firm background in molecular biology or biochemistry who, lured by the media attention that accompanies many ancient DNA results, simply apply the PCR to whatever old specimen they happen to be interested in. They practice what we in the lab liked to privately call "molecular biology without a license."

As I now considered what projects to embark upon in my new lab, I was particularly inclined to study human history by molecular means. It was a fascinating topic but, as generally practiced, riddled with conjecture and biases stemming from preconceived ideas about history. I longed to bring a new rigor to the study of human history by investigating DNA sequence variation in ancient humans. One obvious possibility was to study the Bronze Age humans that were preserved in the peat bogs of Denmark and Northern Germany. But as I read more about them, I realized that these corpses had been preserved because the acid conditions in the bogs had essentially tanned them. Acid conditions lead to nucleotide loss and strand breakage and are therefore extremely bad for DNA preservation. But even worse, the tendency to find human DNA even in animal remains suggested that working with ancient humans could be seriously problematic.

So instead we started to collect samples of extinct animals, such as Siberian mammoths. And we started to do controlled experiments in a systematic way. For example, my first graduate students, Oliva Handt and Matthias Höss, used primers specific for human mtDNA. To my dismay, they found that they could amplify human DNA from almost all our animal samples and generally also from the blank extracts. We made up new reagents from fresh containers that had just been delivered to the lab, but it didn't help. We did this again and again, trying to be as meticulous as we possibly could, but month after month we continued to find human DNA in almost every experiment. I began to despair. How could we ever trust the data, unless they completely conformed to our expectations, such as finding marsupial-like sequences from a marsupial wolf? And if we could only trust the expected results, that would make the field of ancient DNA very boring indeed, as we could then never discover the unexpected—which is, of course, the essence of experimental work and the dream of every scientist.

I walked home night after night frustrated and impatient with our failed experiments. But gradually it dawned on me that I was still being naïve about the contamination issue. I had not drawn the logical conclusions from my awareness of the PCR's extreme sensitivity. At Berkeley, and during the first period in Munich, we would extract DNA from museum specimens on our lab benches—the same benches where we handled large amounts of DNA from humans and other organisms we were interested in. If even a microscopic droplet of a modern DNA solution made it into the ancient DNA extract, the modern DNA would overwhelm the few ancient molecules that might have come from the ancient tissue. This could well happen even if we made no obvious mistakes, such as forgetting to change the plastic tip of a pipette.

It became clear to me that what we needed was to achieve complete physical separation of the extraction and handling of DNA from ancient

FIGURE 4.1. Oliva and Matthias in the first "clean room" in Munich. Photo: University of Munich.

tissues and all other experiments in the lab. In particular, we needed to isolate these experiments from the PCR, where trillions of molecules were produced. We needed a laboratory dedicated solely to ancient DNA extraction and amplification. So we located a small windowless room on our floor, which we emptied out completely and repainted, then spent time thinking about how DNA that might be lurking on the new benches and instruments we bought for this lab could best be destroyed. We came up with some harsh treatments. We cleaned the entire lab with bleach, which oxidizes DNA. We mounted ultraviolet lamps in the ceiling and left them on all night, since UV light wreaks havoc on DNA molecules. And we bought new reagents for our new lab, the first "clean room" in the world devoted to work on ancient DNA (see Figure 4.1). These measures dramatically improved things. Our blank extracts became clean, while, to my delight, some of our samples continued to yield DNA. But gradually, over months, the blanks turned up with DNA again. I was furious. What was going on? We threw out all our reagents and bought new ones.

Things got better again, but only for a while. It was time for paranoia, and the paranoia led not only to my mania for cleanliness in the clean room but to my establishment of several firm rules for how to work in a clean room—rules that to this day remain the standard. First of all, access was limited to the select group who did experiments there—in this case, my first two graduate students, Oliva and Matthias. Before they entered the clean room, they each donned a special lab coat, hairnet, special shoes, gloves, and a face shield. After some additional frustrations with contaminated blank extracts, I decided that they were allowed into the clean room only when they came directly from home in the morning. If they first walked through rooms where PCR products might be present, they were banned from the clean room for the rest of the workday. All chemicals had to be delivered directly to the clean room, and we bought new equipment that also went directly there. Slowly, things got better. Still, all new solutions and chemicals needed to be tested by the PCR for traces of human DNA, and it was not uncommon for a batch to have to be discarded. All of this was taxing work for Oliva and Matthias, who had joined me in hopes of studying ancient humans and extinct animals and found themselves instead vetting chemicals and fretting about contagion.

But the lab's general mood improved as our efforts began to pay off. As our extracts became clean, we could start working on other methodological issues. So far, all our work had been on soft tissues, such as skin and muscle. But I remembered that one of my DNA-yielding mummy samples in Uppsala had come from cartilage, a tissue not very different from bone. If DNA could be extracted from ancient bones rather than just soft tissues, this would obviously open up great opportunities, as bones are what generally remain from ancient individuals. In 1991, Erika Hagelberg and J. B. Clegg of Oxford University had published a paper describing the extraction of DNA from ancient human and animal bones.[5] So when the contamination issue was under control, Matthias tried many methods for getting DNA out of bones, focusing on animals where the risk of contamination was much smaller (as DNA from most animals was rare in our laboratory). Among them was a protocol described in the literature for DNA extraction from microorganisms. It relied on the fact that DNA binds to silica particles—essentially a very fine glass powder—in solutions that contain high salt concentrations. The silica particles could then be thoroughly washed to get rid of all kinds of unknown components that were in many of the samples and that could interfere with the PCR. Finally, the DNA could be released from the silica particles by lowering the salt concentration. This extraction procedure was an arduous process, but it worked and so represented a major step forward.

Matthias and I published the silica extraction method in 1993; the experiment used Pleistocene horse bones, and the mtDNA sequence they yielded was proof that we could retrieve DNA from bones that were 25,000 years old—the first time that reliable DNA sequences from before the last Ice Age were presented.[6] With small modifications, this is still the extraction protocol used in most ancient DNA extractions today. The many frustrations that preceded this paper were evident from our opening remark that our young field was "marred by problems." But this was slowly changing. In fact, without realizing it at the time, Matthias and Oliva had laid the foundations for much of what was to come in the next few years. In 1994, Matthias retrieved the first DNA sequences from Siberian mammoths, working with four individuals between 9,700 and more than 50,000 years old. We sent this work to *Nature*, where it was published together with similar results from Erika Hagelberg, who had isolated DNA from the bones of two mammoths.[7] Although these mtDNA sequences were very short, they hinted at what would be possible if more sequences could be retrieved. We saw, for example, that there were many differences among the DNA sequences from the four mammoths. So we could imagine not only clarifying the relationship of mammoths to the two living members of the same order—the Indian and African elephants—but also tracing the history of mammoths in the Late Pleistocene and on up to their extinction some 4,000 years ago. Things were finally looking brighter for ancient DNA.

This was also a time when our skills in extracting DNA and doing the PCR were applied to other, rather less conventional biological materials. Felix Knauer, a wildlife biologist at the university, showed up one day in my office and asked about the application of our DNA techniques to "conservation genetics," the field that tries to apply genetics to the question of how best to protect endangered species. Felix had collected feces from the last surviving wild population of Italian bears, who lived on the southern slopes of the Alps. I invited Felix and a few other students to try our silica extraction method and PCR from the bear feces. We showed that we could amplify bear mtDNA from such droppings. Previously, the only way to get DNA from an animal in the wild was either to kill it or to shoot it with a tranquilizing dart and draw blood, a risky (and for the animal obviously very disturbing) procedure. We could now study the genetic relationship of the Italian bears to other European bear populations without bothering the bears at all. We published this work as a small paper in *Nature*, in which we also showed that we could retrieve DNA from the plants that the bears had eaten and thereby reconstruct aspects of their diet.[8] Extraction of DNA from droppings collected in the wild has since become common practice in wildlife biology and conservation genetics.

As we were painstakingly developing methods to detect and eliminate contamination, we were frustrated by flashy publications in *Nature* and *Science* whose authors, on the surface of things, were much more successful than we were and whose accomplishments dwarfed the scant products of our cumbersome efforts to retrieve DNA sequences "only" a few tens of thousands of years old. This trend had begun in 1990, when I was still at Berkeley. Scientists at UC Irvine published a DNA sequence from leaves of *Magnolia latahensis* that had been found in a Miocene deposit in Clarkia, Idaho, and were 17 million years old.[9] This was a breathtaking achievement, seeming to suggest that one could study DNA evolution on a time scale of millions of years, perhaps even going back to the dinosaurs! But I was skeptical. From what I had learned in Tomas Lindahl's laboratory in 1985, I had concluded that it was possible for DNA fragments to survive for thousands of years, but millions seemed out of the question. Allan Wilson and I did some simple extrapolations, based on Lindahl's work, in which we determined how long DNA would survive if water were present and conditions were neither too hot nor too cold, neither too acid nor too basic. We concluded that after some tens of thousands of years—and per-haps, under extraordinary circumstances, a few hundreds of thousands of years—the last molecules would be gone. But who knew? Perhaps there was something very special about those fossil beds in Idaho. Before go-ing to Germany, I visited the site. The deposits were formed of dark clay, which was removed by a bulldozer. Upon being pried open, the blocks of clay revealed green magnolia leaves, which rapidly turned black when ex-posed to air. I collected many of these leaves and brought them with me to Munich. In my new lab, I tried extracting DNA from the leaves and found that they contained many long DNA fragments. But I could amplify no plant DNA by PCR. Suspecting that the long DNA was from bacteria, I tried primers for bacterial DNA instead, and was immediately successful. Obviously, bacteria had been growing in the clay. The only reasonable ex-planation was that the Irvine group, who worked on plant genes and did not use a separate "clean lab" for their ancient work, had amplified some contaminating DNA and thought it came from the fossil leaves. In 1991, Allan and I published our theoretical calculations in an article about the stability of DNA,[10] and in a second paper we described my failed attempts to get DNA from the plant fossils from Idaho.[11] This was a sad time, since Allan had fallen severely ill with leukemia the year before. Nevertheless, he made substantial contributions to both papers. He died in July of that year at the young age of fifty-six.

Naïve as always, I thought our paper pointing out the chemical im-possibility of DNA survival over millions of years would halt the search for such super-old DNA. But rather than being the end of things, the Idaho plant fossils were the beginning of a whole new area of research. The next super-old DNAs to pop up were found in amber. Amber is resin that was exuded from trees millions of years ago and solidified into translucent golden clumps; it is found in large quantities in quarries in the Domini-can Republic and on the shores of the Baltic Sea, among other places. Not uncommonly, insects, leaves, and even small animals such as tree frogs can be found entombed in resin. Such inclusions often preserve exquisite details of organisms that lived millions of years ago, and many investi-gators hoped that the same would be true for their DNA. One of the first such papers came in 1992, when a group at the American Museum of Nat-ural History published a paper in *Science* presenting DNA sequences from a 30-million-year-old termite encased in a piece of Dominican amber.[12] This was followed in 1993 by a whole series of papers from a lab headed by

Raul Cano at California Polytechnic State University, in San Luis Obispo, including one on DNA from a weevil between 120 million and 135 million years old found in Lebanese amber[13] and another on DNA from a 35- to 40-million-year-old leaf from the Dominican tree that produced the amber.[14] Cano went on to found a company that claims to have isolated more than twelve hundred organisms from amber, including nine ancient strains of live yeast. Leaving such outlandish claims aside, it seemed to me that one could not rule out the possibility that DNA might be preserved for an extraordinarily long time in amber, since it was probably protected from water and oxygen, the two factors most destructive to the chemistry of DNA. That supposition, however, didn't necessarily mean protection from the effects of background radiation over millions of years, nor did it explain why we had struggled so mightily to amplify traces of DNA a thousand times younger.

The opportunity to find out came in 1994, when Hendrik Poinar joined our lab. Hendrik was a jovial Californian and the son of George Poinar, then a professor at Berkeley and a well-respected expert on amber and the creatures found in it. Hendrik had published some of the amber DNA sequences with Raul Cano, and his father had access to the best amber in the world. Hendrik came to Munich and went to work in our new clean room. But he could not repeat what had been done in San Luis Obispo. In fact, as long as his blank extracts were clean, he got no DNA sequences at all out of the amber—regardless of whether he tried insects or plants. I grew more and more skeptical, and I was in good company. In 1993, Tomas Lindahl, who had been interested in ancient DNA ever since my 1985 visit to his lab, published a highly influential review on DNA stability and decay in *Nature,* in which he devoted a section to ancient DNA.[15] He pointed out—as I had with Allan earlier—that survival beyond a few hundred thousand years was unlikely. He left open the possibility that DNA from specimens encased in amber was an exception; in the meantime, however, I had given up even on the amber.

Tomas had also found the perfect term for super-old DNA: *antediluvian DNA.* We loved it, applied it, and it stuck. But this ridicule could not, of course, deter the enthusiasts. The inevitable happened in 1994, when Scott Woodward of Brigham Young University in Utah published DNA sequences that he and his colleagues had extracted from 80-million-year-old bone fragments—bone that "likely" came from a dinosaur or dinosaurs.[16] Not unexpectedly, this paper appeared in one of the two journals that compete for headline-worthy work and enjoy an often undeserved scientific prestige. This time it was *Science*. The authors had determined many different mtDNA sequences from the bone fragments, and some of them seemed to the authors to be as distant from birds and reptiles as from mammals. They suggested that these might be dinosaur DNA sequences. This seemed ludicrous to me. Thoroughly frustrated by the way the field had developed, Hans Zischler, a meticulous, even slightly pedantic, postdoc in my lab, decided to go after this particular piece of work. When we did a more rigorous analysis of the DNA sequences that the Utah group had published, they seemed closer to mammalian—and indeed human—mtDNA than to birds or reptiles.

Still, they didn't quite seem to be human mtDNA. Explaining what they were takes a bit more explanation of the nature of mtDNA. Recall that mitochondrial genomes are circular DNA molecules of 16,500 nucleotides that reside in mitochondria, organelles located outside the cell nucleus in almost all animal cells. These organelles, as well as their genomes, derive originally from bacteria that almost 2 billion years ago entered primordial animal cells and were hijacked by those cells to produce energy. Over time, the hijacked bacteria transferred most of their DNA to the cell nucleus, where the DNA became integrated into the major part of the genome, situated on chromosomes. Even today in the human germ line, when eggs and sperm cells are formed, a mitochondrion will occasionally break, and fragments of its DNA will end up in the cell nucleus. There, efficient repair mechanisms recognize the ends of broken DNA strands and join them to other DNA ends that may exist if the nuclear genome also happens to carry a break. Thus, now and again, pieces of mtDNA become integrated in our nuclear genome, where, without having any function, they are passed on to new generations. Each of us carries hundreds if not thousands of such misplaced mitochondrial DNA fragments in our cell nuclei that have integrated into our genome at various times in the past. These fragments have different degrees of similarity to our real mitochondrial mtDNA; although they resemble ancestral mtDNA sequences, they have accumulated mutations, unconstrained as they are by any functional requirements in their new life as genetic garbage embedded in nuclear DNA. Hans Zischler had worked in our lab on detecting such integrations of mtDNA into the nuclear genome, and as we considered the putative dinosaur DNA, we wondered whether such mtDNA fragments might be what the Utah group had found. Indeed, given our experience with contaminating human DNA, it seemed probable to us that they had found nuclear versions of human mtDNA with unusual mutations. We decided to look in the human nuclear genome for the sequences they had published. The problem with our plan was that any normal preparation of DNA from human cells ended up containing a mix of both nuclear and mtDNA, and the hundreds and thousands of copies of real mtDNA in the mitochondria of most cells would get in the way of our attempts to detect any mtDNA segments that had left the mitochondrion and settled among the nuclear DNA. Here we were helped by biology. As noted in Chapter 1, we inherit our mtDNA exclusively from our mothers, via the egg, and get no mtDNA from our fathers. This is because the heads of the sperm, which penetrate the egg, contain no mitochondria. So to get nuclear DNA without accompanying mtDNA, our lab simply needed to isolate sperm heads.

I talked to my male graduate students, and there was enough enthusiasm for our work that we all went our separate ways one morning and generated sperm, from which Hans carefully isolated the heads by centrifugation. He then purified the DNA from the sperm heads and used the same primers for the PCR as had been used by the Utah group. As expected, he obtained many sequences from nuclear mtDNA fragments, which we then sifted through for any similarity to the "dinosaur" sequences from Utah. Indeed, we found two that were almost identical to the published sequences. This meant that instead of dinosaur DNA, the Utah group had sequenced bits of translocated human mtDNA from the human nuclear genome. Because these segments had left the human mtDNA genome in the distant past, they had picked up enough mutations to appear somewhat distant from humans, yet still similar to mtDNAs from mammals, birds, and reptiles. I could not prevent myself from being slightly facetious when writing the "Technical Comment" for Science[17] and suggesting that there were three possibilities to explain how we could obtain DNA sequences very similar to the ones from Utah using our own DNA in our lab. The first was that we had contamination in our laboratory from dinosaur DNA, which I suggested was unlikely. The second was that dinosaurs hybridized with early mammals before becoming extinct some 65 million years ago. This alternative, too, was dismissed as unlikely. The third (and most plausible) scenario was contamination by human DNA in the dinosaur experiments. Science published our comment along with comments from two other groups, both of which pointed out deficiencies in the DNA-sequence comparisons that had led the Utah group to claim that the mtDNA sequences looked ancestral to birds.

The comment was fun to write but also somewhat bitter, given that studies such as the Utah one had become a constant feature of the ancient

DNA field. The problem of high-profile but dubious results still plagues research on ancient DNA today. As my students and postdocs have often remarked to me, it is easy to generate outlandish results with the PCR but difficult to show that they are correct; nevertheless, once results are published, it is even more difficult to show that they are wrong and explain where the contamination came from. In this instance, we were successful, but our efforts involved a lot of work and did not take our research forward. To this day, it is unclear just where the amber sequences that were published in *Nature* and *Science* came from. With enough work, I was sure the sources could be found, yet I decided we had had enough. As one student put it, "Let's stop playing the PCR police." We determined from then on to ignore those reports we thought were wrong and concentrate on our own work. Our best service to the field, we felt, was to establish methods to retrieve DNA from sources that were some tens of thousands of years old and to show that the results were genuine and correct. With ancient remains of humans this was hard if not impossible, as modern human DNA was lurking almost everywhere. So even though it pained me, I needed to forget about human history for the time being and divert my work to ancient animals. After all, I was a professor in a zoology department. I decided that we would focus on questions concerning the relationships of extinct animals and their present-day relatives.

Chapter 5
Human Frustrations

During his collecting expeditions in the 1830s in South America, Charles Darwin was fascinated but puzzled by fossil remains of various large, plant-eating mammals. These creatures seemed much bigger than any animals currently living in the area. Along with examples of every living animal and bird he could capture, Darwin collected a number of fossils to send back to England, including a large lower jawbone that was eroding out of a coastal cliff in Argentina. The anatomist Richard Owen analyzed the jaw and attributed it to a giant ground sloth the size of a hippopotamus, which he dubbed *Mylodon darwinii* (see Figure 5.1). Even more interesting than the idea of such a bizarrely large herbivore was the idea that it might even still exist, alive, somewhere in the wilds of Patagonia. In 1900, the sensational discovery of apparently fresh dung and skin remnants of what appeared to be giant ground sloths motivated an expedition by a Mr. Hesketh Prichard in search of this marvel. After journeying some two thousand miles through Patagonia, Prichard briskly concluded that he had found "no single scrap of evidence of any kind which would support the idea of the survival of the Mylodon."[1] This was with good reason: we now know that it became extinct during the last Ice Age, some 10,000 years ago.

Two- and three-toed sloths exist today in South America, but, weighing in at a mere ten to twenty pounds, they are of modest size compared to *Mylodon*. And unlike *Mylodon*, both two- and three-toed sloths live in trees. But they seem to have adapted to life in the trees only rather recently in evolutionary terms, since they are rather large for tree-dwelling mammals, not particularly agile aloft, and prefer to descend to the ground for such mundane routines as defecation. A big question was whether the ancestors of the tree sloths had become adapted to the arboreal lifestyle just once, and not particularly gracefully, or whether the two forms of tree sloths were examples of parallel adaptations, whereby ground-dwelling sloths in the past had at least twice independently taken to the trees. If similar adaptations

FIGURE 5.1. Reconstruction of ground sloth skeleton. Source: http://commons .wikimedia.org/wiki/.

happened independently more than once—if history repeated itself, so to speak—it suggests that there are a limited number of ways in which animals can adapt to an ecological challenge. Each such case of convergence, when two or more unrelated organisms independently evolve similar behaviors or body shapes, is evidence that evolution follows rules—and is helpful in deducing how these rules work. An example of this was the marsupial wolf that we had studied in Zurich and Berkeley. In the case of the tree sloths, just as in the case of the marsupial wolf, we could determine whether convergence had occurred if we could clarify how Darwin's extinct giant ground sloth was related to the two-toed and three-toed tree sloths.

I visited the Natural History Museum in London and spent some time there with the amiable curator of Quaternary mammals, Andrew Currant, an expert on mammal paleontology with a build not unlike that of a large Pleistocene mammal. He showed me some of the fossilized bones that Darwin had brought back, and he allowed me to cut a small piece from two of the Patagonian *Mylodon* bones in their collection. I also visited the American Museum of Natural History, in New York, and got samples for our study there. But it was in Andrew's museum that I experienced a vivid demonstration of how readily the ancient animal specimens we studied might become contaminated. As I was examining sloth bones with Andrew, I asked him if they had perhaps been treated with varnish. To my amazement, he picked up a bone and licked it. "No," he said, "these have not been treated," explaining that if a bone had been treated with varnish, it would not absorb saliva. In contrast, an untreated bone would do this so efficiently that one's tongue tended to adhere to the bone. I was horrified and wondered how many times this "test" had been done during the hundred years or more that some of the bones we worked with had been in museums.

Once the samples were back in Munich, Matthias Höss applied his skills to them. As always, I insisted that we first pay attention to the technical side of things. My interest in sloths was after all driven mainly by an interest in how to retrieve ancient DNA. Matthias used a rough assay to estimate the total amount of DNA in his *Mylodon* extract and another crude assay to measure how much of that was similar to modern sloth DNA. It turned out that about 0.1 percent of the DNA in our best *Mylodon* bone extract was from the animal itself, the rest having come from other organisms that had lived in the bones after the giant sloth died. This has turned out to be typical of many ancient remains we have since studied.

Focusing on mitochondrial DNA fragments, Matthias managed to use the PCR to reconstruct a stretch of *Mylodon* mtDNA more than a thousand nucleotides long by amplifying short overlapping pieces. By determining and comparing the same sequences from samples from living sloths, he could show that the giant ground sloth, which stood ten feet tall on its hind legs, was more closely related to the present-day two-toed tree sloth than to the three-toed tree sloth. This was important, since if the two- and three-toed sloths had been most closely related to each other and more distantly related to *Mylodon* (which was the opinion of most scientists at that time), it would have suggested that they had a common ancestor who became tree-dwelling. Our result suggested that sloths had at least twice evolved into forms that were small and spent most of their lives in trees (see Figure 5.2).

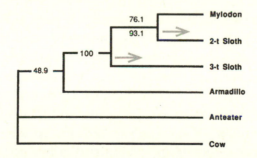

FIGURE 5.2. A tree showing that the Mylodon is more closely related to the two-toed than to the three-toed sloth, suggesting that sloths started to live in the trees twice during their history. From Matthias Höss et al., "Molecular phylogeny of the extinct ground sloth *Mylodon darwinii*," *Proceedings of the National Academy of Sciences USA* 93, 181–185 (1996).

That both the marsupial wolf and the tree sloths turned out to be examples of convergent evolution was to me a strong message that morphology is often an unreliable indicator of relatedness among organisms. It seemed that almost any body shape or behavior could evolve independently if a change of environment created a pressure for a change in lifestyle. To me, DNA sequences seemed to offer a much better chance to correctly gauge how species were related to each other. DNA sequences can accumulate hundreds and thousands of mutations over time, each of which occurs independently of one another, and most of which have no influence on how an organism looks or behaves. In contrast, measurement of morphological features necessarily is done on traits that might very well affect the survival of the organism, and the sizes of different features, such as various bones, might be linked to one another. Because of the greater number of independent, randomly varying data points that can be accumulated, DNA sequences allow reconstruction of relationships with greater rigor than morphological features. In fact, in contrast to morphological features, even the timing of divergences from a common ancestor can be derived from the number of differences that have accumulated in DNA sequences, since these differences occur roughly as a function of time, at least within a group of related animals.

Matthias used such a "molecular clock" approach and calculated the numbers of nucleotide differences and underlying mutations that had accumulated in the mtDNAs of members of the group of animals to which the sloths belong, which includes armadillos and anteaters. He found that this group of animals is surprisingly old. They began diversifying before the dinosaurs became extinct, some 65 million years ago. This timetable holds for some other groups of mammals, as well as for birds, such that many groups of present-day animals trace back to ancestors that originated at a time when dinosaurs dominated the earth. Once, many different forms of ground-dwelling sloths existed, but today there are only tree sloths. Until our discovery that today's tree sloths do not have a common ancestor, it had been reasonable to think that the arboreal forms might share some unknown but important physiological adaptation that allowed them to survive, perhaps in the face of climatic change during the last glaciation. But if they did not share a common ancestor, this seemed less likely. It was more plausible that the crucial factor in their survival was the most obvious one: that they live in trees. We ended our paper with the speculation that living in trees might have helped them survive the arrival of humans, who seem to have hunted ground-dwelling, slow-moving sloths to extinction.[2] Although the debate continues about whether ecological factors or overhunting by humans caused the disappearance of American megafauna, such as ground-dwelling sloths, by approximately 10,000 years ago, we were happy that ancient DNA could add a piece to the puzzle. We had shown that reliable DNA sequences could be retrieved from animals that lived thousands of years ago, and that this could yield enough information to provide a new perspective on their evolution.

By the mid-1990s, the field of ancient DNA studies had somewhat stabilized. Many researchers had come to realize what was possible and what was not. Zoological collections of skins and other parts of animals that had been dried shortly after death could be used routinely for DNA extraction, as we had shown by our investigations of kangaroo rats, back in Berkeley. Studies on pocket gophers, rabbits, and many other animals followed, and several of the big zoological museums established molecular labs in the 1990s devoted to the DNA study both of their old collections and of new samples collected specifically for that purpose. The Smithsonian Institution in Washington, DC, and the Natural History Museum in London were among the first to do this, and they were followed by others. Similarly, forensic scientists analyzed DNA that they could now extract and amplify from evidence collected many years earlier. This led to courts overturning the convictions of wrongfully imprisoned people and to new strides, based on genetic evidence, in identifying remains and apprehending criminals. The despondency of my first years in Munich, when I and my group had struggled with contamination and other methodological issues while others published nonsensical millions-of-years-old DNA sequences in *Science* and *Nature,* was now replaced by a sense of satisfaction that it had all been worthwhile. The field had become established. It was time to return to the old challenge: human remains.

As noted, there are plenty of ways that modern human DNA can contaminate an experiment. The curator in London had demonstrated one very obvious way to me when he put his tongue to the sloth bone, but dust, bad reagents, and much else also pose problems. Human history was, for me, the ultimate goal. The question was whether, despite the impediments, we could find a way forward.

Oliva Handt devoted herself to this quest. Oliva is a warm, almost motherly person who tended to be overcritical of her own work. I felt that this was an excellent trait for the job she was about to undertake. She had to deal with the same issues as Matthias did in his sloth work, but in addition she had to worry about the odd dust particle that might land in a test tube where the extract of an ancient human bone was kept. If no such dust particle had landed in the blank extract performed in parallel with the bone extract, then it might be hard or impossible to tell whether or not the sequence she determined was from the bone or from a contaminating dust particle. For this reason, we decided that Oliva would work on Native American remains, whose mtDNA contained certain variants not found in Europeans. Although I much disliked doing experiments where only results in accordance with pre-expectations would be credible, this seemed to be one of the few ways in which we could reliably work out the methods for retrieving ancient human DNA sequences. So Oliva started working on skeletons and mummified human remains from the American Southwest that were around 600 years old. As she was toiling away on this, repeating her extractions over and over to test the reproducibility of her results, there came an opportunity that was too good to pass up.

In September 1991, two German hikers had found the mummified body of a man in the Alps in the Oetztal near Hauslabjoch, on the border between Austria and Italy. They, and the authorities they contacted, first believed it to be a modern corpse, perhaps a war victim or an unfortunate hiker who had been lost in a snowstorm. But after the man's body was removed from the ice, what was left of his clothes and equipment made it plain that he was not a recent soldier or hiker; rather, he died on the Alpine pass about 5,300 years earlier, during the Copper Age. From the news media, I learned that the Austrian and Italian governments were each claiming that the mummy had been found on their territory. There were also disputes between the discoverers and government officials about finders' fees, and difficulties with the people in the pathology department of the University of Innsbruck, in Austria, who kept the frozen corpse and jealously guarded him against outsiders. In short, it seemed to be a legal and general mess. I was therefore surprised when, in 1993, I was approached by a professor from Innsbruck who asked if we wanted to analyze the DNA of the Ice Man—or Oetzi, as he had come to be called, after the valley in which he was found. We expected that a body that had been frozen continuously for more than 5,000 years would be much better preserved than any mummies from Egypt or bones from North America. We decided to give it a try.

Oliva and I traveled to Innsbruck, where the pathologists removed eight small samples from Oetzi's left hip, which had been damaged when the body (as yet unrecognized as being ancient and unique) was freed from the Alpine ice by means of a sledgehammer. Back in Munich, Oliva set about extracting and amplifying mtDNA. We were all excited when she obtained nice PCR products, but when she sequenced them, the sequences were uninterpretable. At many positions, there seemed to be several different nucleotides present. To sort this out, she went back to the old cloning approach I had used in Uppsala: she cloned each PCR product and then sequenced several clones. Since each clone came from a single original DNA fragment that had been amplified by the PCR, she could see whether all the original DNA fragments carried the same nucleotide sequence as would be expected if they came from a single individual or, alternatively, whether they carried different sequences and therefore came from different individuals. The latter turned out to be the case, and in fact different samples gave different mixtures of sequences. This was maddening. Most, if not all, of the mtDNA must have come from people who had handled the Ice Man since he was discovered. How were we to determine whether a sequence came from the Ice Man or not? After all, in evolutionary terms, he had not lived all that long ago, so he would undoubtedly have a version of mtDNA similar or identical to those found in Europeans today, and many Europeans had apparently come into contact with him since his discovery.

Fortunately, two of the samples we had obtained in Innsbruck were large enough for us to remove the surface tissue and extract material from the inside, untouched part of the sample, in hopes that whatever contamination was present would primarily be on the surface. This helped, but only to a degree. Oliva found six positions where she saw a mixture of sequences suggesting that the number of different mtDNA variants was smaller, coming from perhaps three or four individuals. But the sequences did not neatly group into three or four classes of identical sequences. Oliva found that the variants at the six positions were scrambled among the molecules, especially when she looked at positions far removed from one another. This must be the result of the "jumping PCR" I had described in Berkeley, whereby instead of copying a single continuous piece of DNA, the polymerase stitches fragments of DNA together into new combinations. Could this scrambled mixture of DNA sequences be disentangled so that we could decide which, if any, of these sequences was from the Ice Man?

We argued that the jumping phenomenon should occur primarily in attempts to amplify longer pieces of DNA rather than shorter ones because shorter pieces would be more likely to be preserved in an intact form in the tissues while longer ones were likelier to be stitched-together molecules or contaminants. So Oliva did a PCR of extremely short pieces. This helped. Whenever she amplified pieces shorter than about 150 nucleotides, not only did she not get scrambled sequences but almost all of her clones carried the same sequence. The picture was becoming clearer. Our extracts contained one mtDNA sequence that was there in large quantities but degraded into short pieces. They also contained mtDNA sequences from two or more additional people that were less frequent and present in larger pieces. We suggested that the abundant and more degraded DNA was likely to have come from the Ice Man, whereas the other DNAs, which were less abundant but also less degraded, were likely to have come from modern individuals who had contaminated the Ice Man.

By amplifying each of the short pieces at least twice, cloning them, and sequencing several clones from each amplification product, Oliva could eventually reconstruct the mtDNA sequence that the Ice Man was likely to have carried when alive. The overlapping fragments she generated determined a sequence of a bit more than 300 nucleotides. Only two substitutions distinguished it from a commonly used reference sequence for modern European mtDNA, and the identical sequence is not uncommon in Europe today. This was not terribly unexpected. From the perspective of someone hoping to live for 80 or 90 years, 5,300 years is a long time, representing about 250 generations. From an evolutionary perspective, however, this is a short time. Unless major disasters, such as epidemics, kill much of a population or major population replacements occur, not much would change in our genes in 250 generations. Indeed, my lab mates and I had anticipated that at most one mutation would have occurred in the segment we studied since the Copper Age.

But before we could publish our results, we faced one additional hurdle: our rule, borne out of frustration with the many unreliable results published in the field, that important or unexpected results should be replicated in a second laboratory. The sequence we had determined from the Ice Man was not biologically unexpected but it certainly would attract attention, so this was an opportunity to show how things should be done. We decided to send one of our unused tissue samples to Oxford, where Bryan Sykes, a geneticist who had left his earlier career in connective-tissue diseases to work on mtDNA variation in humans and in ancient DNA, was eager to help. Sykes's student extracted and amplified a piece of the sequence we had determined and reported the sequence back to us. It was identical to Oliva's, and we described our findings in a paper in *Science*.[3]

Although regarded as a success at the time, to my mind this experience primarily showed how difficult it was to work with ancient human remains. The Ice Man had been frozen and could be expected to be unusually well preserved, and moreover had been found only two years earlier, so that not too many people had had the chance to contaminate it, yet we had found a mixture of different sequences that had been difficult to sort out. We had succeeded only thanks to Oliva's patience and perseverance and our inference of what was likely to be the correct sequence, which necessarily relied on assumptions about the different populations of molecules in the tissue. Any study of recent human evolution, where one would need to study populations of many individuals, probably all preserved as skeletons, seemed too daunting to contemplate.

On the bright side, we had gained an enormous amount of experience with human sample materials and a better appreciation of the difficulties involved. To profit from this, Oliva returned to the Native American remains. As we expected, it wasn't easy. My friend Ryk Ward arranged for us to get ten samples of mummies that were about 600 years old from Arizona in the American Southwest. As one might imagine, the outcome was comparable to that of the Ice Man analysis. From nine of the individuals, Oliva either could not amplify anything at all or found sequences so mixed up that it was impossible for her to determine whether any of them was endogenous to the individual. In only one case was she able to do short amplifications, and by sequencing many clones from repeated amplifications she showed that this sample contained relatively many molecules and that they came from an mtDNA that resembled mtDNA sequences found among modern Native Americans. Somewhat frustrated, we wrote, in the summary of a 1996 paper describing Oliva's work, that "these results show that more experimental work than is often applied is necessary to ensure that DNA sequences amplified from ancient human remains are authentic."[4] This, obviously, was also an implicit criticism of much of the work that was being performed by others on ancient human remains.

Despite all of Oliva's efforts, I now decided to abandon all work on ancient human remains. Other laboratories continued to publish results, but I felt that much of what appeared was unreliable. The situation was deeply frustrating.[5] In 1986, I had left what looked like the start of a promising career in medical research because I wanted to introduce new and accurate methods of studying human history in Egypt and elsewhere. By 1996, I had been able to establish reliable methods that turned zoological museums into veritable gene banks and made it possible to study mammoths, ground sloths, ancestral horses, and other animals from the last Ice Age. This was all well and good, but it was not where my heart was, and I worried that I would turn into a zoologist against my will.

I did not torture myself daily with these thoughts, but again and again, when I reflected on what I might do in the future, I felt frustrated. What I *wanted* to do was to illuminate human history, but it seemed next to impossible to study ancient humans because, in most cases, their DNA could not be distinguished from that of living people. But after a while I realized that I could perhaps do something of even greater relevance to understanding human history than the study of DNA from Bronze Age people or Egyptian mummies. Perhaps I could study people of another sort, people who had been in Europe long before the Ice Man—the Neanderthals.

Turning to Neanderthals might seem strange, given that I had just sworn off ancient humans. But of crucial importance for me was that they could be expected to have DNA sequences that were recognizably different from present-day humans. This was not just because they lived more than 30,000 years ago but also because they'd had a long history different from ours. Some paleontologists estimated that we shared a common ancestor with them at least 300,000 years ago, and some said they were a different species. Anatomically, the Neanderthals looked strikingly different from present-day humans and also from early modern humans living elsewhere in Europe at approximately the same times. Yet Neanderthals are the closest evolutionary relatives of all contemporary humans. Studying how we differed genetically from our closest relatives would potentially allow us to find out what changes set apart the ancestors of present-day humans from all other organisms on the planet. In essence, we would be studying perhaps the most fundamental part of human history—the biological origin of fully modern humans, the direct ancestors of all people alive today. Such research might also tell us exactly how Neanderthals were related to us. Neanderthal DNA seemed like the coolest thing imaginable to me. And by sheer luck I was in Germany, where Neander Valley is situated, where the first Neanderthal had been found, the so-called type specimen used to define Neanderthals. I wanted desperately to get in touch with the museum in Bonn where the Neanderthal type specimen was housed. I had no idea whether its curators would be reluctant to give me a sample. This type specimen was, after all, what some (perhaps in an attempt to forget certain aspects of twentieth-century German history) called "the most well-known German." It was something of an unofficial national treasure.

I worried for months over how to go about this. I knew all too well how tricky it could be to work with museum curators, who were entrusted with the difficult task of preserving valuable specimens for future genera-tions while at the same time facilitating research. I had found, in some in-stances, that they considered their primary role to be the exertion of power, refusing access to specimens even when the possible gain of knowledge seemed to far outweigh the value of preserving a small piece of a bone. If such curators were approached in the wrong way, they could say no and then, for all-too-familiar reasons of human pride, find it difficult to go back on their word. While fretting about this, I received one day, in the most remarkable of convergences, a phone call from Bonn. The caller was Ralf Schmitz, a young archaeologist who, together with the Bonn museum cu-rator, was responsible for the Neanderthal type specimen. He asked if I remembered an exchange we had had a few years back.

He reminded me that in 1992 he had asked what the chances of success were if one were to try to obtain DNA from a Neanderthal. This conversation had slipped out of my mind, being one out of many with ar-chaeologists and museum curators. Now I remembered. At the time I had not known what to answer. My immediate, and slightly delinquent, im-pulse had been to suggest that the chances were good, so that they might readily part with a Neanderthal bone. But almost as quickly I had realized that honesty was the best way forward. After some hesitation, I had said that in my opinion we might have a 5 percent chance of success. Ralf had thanked me and I had heard nothing more from him since.

Now, almost four years later, Ralf was on the phone and said that, yes, we would be allowed to have a piece of the Neanderthal from Nean-der Valley. As it turned out (Ralf later told me), others had approached the museum to request samples, saying they were almost certain to get usable DNA out of the specimen. The museum authorities had then prudently decided to get the opinion of another lab and had asked Ralf to contact me. Not only our track record but also our apparent honesty in suggesting that the chances were slim had convinced Ralf and the museum that we would be their best partners. They were, as it turned out, the exact opposite of the obstructionist museum curators I had been fretting about. I was delighted.

What followed were weeks of discussions with the museum about how much bone material we would get, and from which part of the skeleton. In total, there was about half of a skeleton of what seemed to be a male individual. Our experience had shown us that the best chances of success were with compact bone—for example, a part of the shaft of an arm or leg bone, or the root of a tooth, rather than thin bones with a large marrow cavity, such as ribs. Eventually we agreed on a piece of the right upper arm,

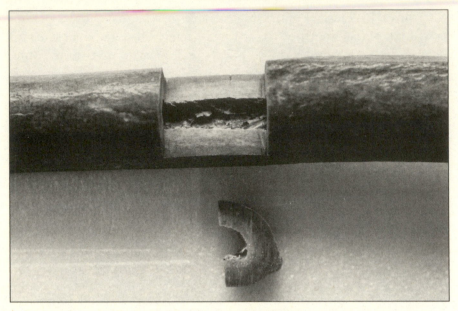

FIGURE 5.3. The right upper arm bone of the Neanderthal type specimen with the sample removed by Ralf Schmitz in 1996. Photo: R. W. Schmitz LVR-LandesMuseum Bonn.

from a part where the shaft had no ridges or other features of interest to paleontologists, who study how muscles had attached to the bone. It also became clear that we would not be allowed to remove the sample ourselves. Ralf and a colleague came to see us in Munich, and we gave them a sterile saw, protective clothing, sterile gloves, and containers in which to store the sample—and off they went. In the end, it was probably fortunate that I was not allowed to put the saw to the archetypal Neanderthal myself. I would probably have been too intimidated by this iconic fossil and would have cut off a very small piece, perhaps too small for success. When we received the sample, we were impressed by the size of what they had removed—3.5 grams of what looked like very well-preserved whitish bone (see Figure 5.3). Ralf reported that when they sawed through the bone, a distinct smell of burnt bone spread through the room. This, we believed, was a good sign; it had to mean that collagen, the protein that makes up the matrix of bone, had been preserved. It was with awe and trepidation that I approached my graduate student Matthias Krings, who had spent more than a year on fruitless attempts to extract DNA from Egyptian mummies—with the plastic bags containing the piece of the Neanderthal type specimen and asked him to apply our latest and best methods to it.

Chapter 6
A Croatian Connection

During the weeks and months after our publication of the Neanderthal mtDNA sequence, I reflected on what had led up to it. I had come a long way from my first attempts sixteen years earlier to extract DNA from a piece of dried calf's liver from the supermarket. Now, for the first time, we had used ancient DNA to say something new and profound about human history. We had shown that the archetypical Neanderthal carried mitochondrial DNA very different from the mtDNA in people today, and that he or his relatives had not, before they became extinct, contributed their mtDNA to modern people. The achievement had required years of painstaking work to develop techniques to reliably determine DNA sequences from individuals long dead. Now that I had these techniques at my disposal, and a group of dedicated people able and willing to try new things, the biggest question was: Where should we go from here?

One task seemed of immediate importance: to determine mitochondrial DNA sequences from other Neanderthals. As long as we had studied only one individual, it remained possible that other Neanderthals carried mitochondrial genomes very different from the one from Neander Valley, perhaps even carrying mitochondrial genomes that were like those of present-day humans. Mitochondrial DNA sequences from additional Neanderthals would also reveal something about the genetic history of the Neanderthals themselves. Present-day humans, for example, have relatively little genetic mtDNA variation. If Neanderthals did, too, this would suggest that they had originated and expanded from a small population. If, on the other hand, they had as much mtDNA variation as any of the great apes have, this would suggest that over their history their numbers had never been very low. They would not have had such a dramatic history with ups and downs in population size as modern humans have. Matthias Krings, eager to follow up on his success with the iconic type specimen from Neander Valley, was keen to examine other Neanderthal specimens.

The major problem was getting access to fossils sufficiently well preserved for us to do work.

I thought a great deal about why we had been successful with the Neander Valley type specimen and came to realize that the fact that it had come from a limestone cave might be significant. Tomas Lindahl had taught me that acid conditions cause DNA strands to disintegrate, which was why the Bronze Age people found in acid bogs in northern Europe had never yielded any DNA. But when water passes over limestone, it becomes slightly alkaline. So I decided we should concentrate on Neanderthal remains unearthed in limestone caves.

Unfortunately, I had never paid much attention to the geological features of Europe in school. But I remembered the first anthropological conference I had ever attended, in Zagreb, in what was then Yugoslavia, in 1986. During the conference, we were taken on excursions to Krapina and Vindija, two sites where large amounts of Neanderthal bones had been found in caves. I made a quick search in the literature and confirmed that both Krapina and Vindija were limestone caves, which was promising. Promising as well was the presence of large numbers of animal bones, particularly of cave bears, in the caves. Cave bears, which were a large plant-eating species, became extinct shortly after 30,000 years ago, just like the Neanderthals. Their bones often abound in caves, often in circumstances suggesting that they died during hibernation. I was happy about the presence of cave-bear bones because they could possibly serve as a convenient tool to check whether DNA was preserved in the caves. If we could show that their bones contained DNA, this might be a good means to convince hesitant curators that they should allow us to try the much more valuable Neanderthal remains from the same cave. I decided to interest myself in cave-bear history, especially in the Balkans.

After a bloody war with Serbia, Zagreb had become the capital of the independent Republic of Croatia. The largest collection of Neanderthals there is from Krapina, in northern Croatia, where starting in 1899 the paleontologist Dragutin Gorjanović-Kramberger discovered more than eight hundred bones from some seventy-five Neanderthals—the richest cache of Neanderthals ever found. These bones are today housed in the Museum of Natural History in the medieval center of Zagreb. The other site, Vindija Cave (see Figure 6.1) in northwestern Croatia, was excavated by another Croatian paleontologist, Mirko Malez, in the late 1970s and early 1980s. He

FIGURE 6.1. Vindija Cave in Croatia. Photo: J. Krause, MPI-EVA.

found bone fragments of several Neanderthals but no spectacular crania like those found in Krapina. Malez also found enormous amounts of cave-bear bones. His finds are housed in Zagreb, too, in the Institute for Quaternary Paleontology and Geology, which belongs to the Croatian Academy of Sciences and Arts. I arranged to visit both this institute and the Museum of Natural History. In August 1999, I arrived in Zagreb.

The Krapina Neanderthal collection was extremely impressive, but I was skeptical about its potential for DNA research. The bones were at least 120,000 years old and therefore older than anything we had found to yield DNA. The Vindija collection looked more promising. First of all, it was younger. Several layers in the excavation had yielded Neanderthal remains, but the uppermost and thus the youngest one to do so was between 30,000 or 40,000 years old—young, as far as Neanderthals go. I saw a second exciting feature of the Vindija collection: it was full to overflowing with ancient cave-bear bones. They were stored, according to bone type and layer, in innumerable paper sacks that were coming apart in the humidity of the Quaternary Institute's basement. There were sacks full of ribs, others full of vertebrae, others of long bones, and yet others of foot bones. It was an ancient DNA gold mine.

In charge of the Vindija collection was Maja Paunovic, a woman of a certain age who spent her days in an institute without public exhibitions and with few facilities for doing research. She was friendly enough but understandably dour—no doubt aware that science had passed her by. I spent three days with Maja, going through the bones. She gave me cave-bear bones from several layers at the Vindija site as well as small samples of fifteen of the Neanderthal bones. This seemed exactly what we needed for the next step in our exploration of the genetic variation among Neanderthals. When I flew back to Munich I felt confident that we would make quick progress.

In the meantime, Matthias Krings had extended his sequencing of the Neanderthal type specimen to a second part of the mitochondrial genome. The results confirmed that this specimen's mitochondrial DNA shared a common ancestor with present-day human mtDNAs about half a million years ago. But this was of course what we had expected, so the news felt slightly boring after the emotional high produced by the first Neanderthal sequences. Not surprisingly, he was eager to throw himself on the fifteen Neanderthal bone samples I had gotten from Maja in Zagreb.

We first analyzed their state of amino-acid preservation. Amino acids are the building blocks of proteins and can be analyzed from much smaller samples than are needed for DNA extractions. We had shown before that if we could not find an amino-acid profile suggesting that the samples contained collagen (the main protein in bones), and if the amino acids were not present largely in the chemical form in which they are built into proteins by living cells, then our chances of finding DNA were very small and there would be no point in destroying a larger piece of the bone in an attempt to extract it. Seven of the fifteen bones looked promising, with one that particularly stood out. We sent a piece of that bone for carbon dating and the result showed that it was 42,000 years old. Matthias made five DNA extracts and amplified the two mitochondrial segments he had studied in the type specimen. It worked nicely. He sequenced hundreds of clones, taking pains to ensure that every position was observed in at least two amplifications that, at my insistence, should come from different extracts, in order to make absolutely sure that they were totally independent of each other.

In March 2000, while Matthias was working on this, a paper that appeared in *Nature* took us by surprise. A group based in the UK had sequenced mtDNA from another Neanderthal, unearthed at Mezmaiskaya Cave in the northern Caucasus.[1] They had not applied all the technical approaches we advocated to make sure that the sequence was correct; for example, they had not cloned the PCR products. Nevertheless, the DNA sequence they found was almost identical to our type-specimen sequence from Neander Valley. Matthias, who had his sequences almost finished, was disappointed that he'd been beaten to the publication of the world's second Neanderthal mtDNA sequences—especially since his progress had been slow due to all the precautions and checks on which I insisted. I sympathized with him, but I was also happy that our pioneer sequence from Neander Valley had been verified by a group working independently of us. Yet I did not quite agree with the commentary *Nature* published along with the paper, which said that this second Neanderthal sequence was "more important" than the first because it showed that the first was correct. I wrote that off as sour grapes on *Nature's* part for not getting to publish the first Neanderthal sequence.

There was a consolation prize of sorts for Matthias. The second Neanderthal DNA not only served to confirm the results in our 1997 *Cell* paper, but now that we knew three sequences, including the one Matthias had determined from Vindija, it became possible to say something, albeit something tentative, about genetic variation among Neanderthals. Genetic theory holds that with just three sequences there is a 50 percent chance of sampling the deepest branch of a tree relating all the mitochondrial DNAs in a population. It turned out that 3.7 percent of the nucleotides in the segment that Matthias and the British group had sequenced differed among the three Neanderthals. For perspective, we wanted to compare this degree of variation to the variation in humans and the great apes. First, we used sequence data for the same segment determined by many other groups from 5,530 humans from all over the world. In order to make a fair comparison to the three Neanderthals, we sampled three randomly chosen humans many times, so that we could calculate an average of how many differences three humans carry in the same sequence. It was 3.4 percent, very similar to that for the three Neanderthals. There were 359 chimpanzee sequences available for the same mtDNA segment. When we sampled chimpanzees in the same way, they differed by an average of 14.8 percent, and for twenty-eight gorillas the corresponding value was 18.6 percent. So Neanderthals seemed to be different from the great apes in having little mtDNA variation, just like present-day humans. Obviously, it was risky to speculate from just three individuals, and from just mtDNA, so when we published these data later in 2000, in *Nature Genetics,* we stressed that it would be desirable to analyze more Neanderthals; nevertheless, we suggested that Neanderthals were probably similar to modern humans in having little genetic variation and that they had therefore expanded from a small population, just like us.[2]

Chapter 7
A New Home

Life is not an orderly thing. One morning, not long before our publication of the first Neanderthal mtDNA sequences in 1997, my secretary told me that an elderly professor had phoned asking for an appointment with me. He had told her that he wanted to discuss some plans for the future. I had no idea who he was but vaguely supposed he was a retired professor who wanted to share his crackpot ideas about human evolution with me. I was very wrong. What he had to say was very exciting.

He explained to me that he came on behalf of the Max Planck Society, or MPS for short, which supports basic research in Germany. Among its many efforts, the MPS had a program to build up world-class research in the former East Germany, which had been fused with West Germany seven years earlier. One guiding principle was to found new research institutes focusing on topics in which Germany was scientifically weak. An area of particular weakness was anthropology, and for a very good reason.

As do many contemporary German institutions, the MPS had a predecessor before the war. Its name was the Kaiser Wilhelm Society, and it was founded in 1911. The Kaiser Wilhelm Society had built up and supported institutes around eminent scientists such as Otto Hahn, Albert Einstein, Max Planck, and Werner Heisenberg, scientific giants active at a time when Germany was a scientifically dominant nation. That era came to an abrupt end when Hitler rose to power and the Nazis ousted many of the best scientists because they were Jewish. Although formally independent of the government, the Kaiser Wilhelm Society became part of the German war machine—doing, for example, weapons research. This was not surprising. Even worse was that through its Institute for Anthropology, Human Heredity, and Eugenics the Kaiser Wilhelm Society was actively involved in racial science and the crimes that grew out of that. In that institute, based in Berlin, people like Josef Mengele were scientific assistants while performing experiments on inmates at Auschwitz death camp, many of them children. Whereas Mengele was sentenced for his crimes after the war (although he had escaped to South America), his superiors at the Institute for Anthropology were never charged. On the contrary, some of them became professors at universities.

When the Max Planck Society was formed in 1946 as the successor of the Kaiser Wilhelm Society, anthropology was understandably a subject best avoided. In fact, as a result of what happened under Nazi rule, the entire field of anthropology had lost its status in Germany. It failed to attract funding, good students, and innovative researchers. Obviously, this was an area where Germany was scientifically weak, and my visitor said that the MPS had set up a committee to consider whether anthropology could be an area in which the MPS should found a new institute. He also indicated that there were different opinions as to whether this was a good idea given recent German history. Nevertheless, my visitor asked me if I would consider moving to such an institute should it be created. I was vaguely aware that the MPS commanded great resources, and that these resources had been augmented to build several new institutes in the east after reunification of the two German states. I was intrigued by the prospect of helping build a new institution, but I did not want to sound overenthusiastic and have them believe I would come under any circumstances. With this in mind I said that I would consider it if it meant that I would be able to influence how such an institute would be organized and function. The professor assured me that as a founding director I would have great freedom and influence. He suggested I come and present to the committee my views on how such an institute might be organized.

Some time later I received an invitation to hold a presentation in front of the committee. It would meet in Heidelberg and was made up of several foreign experts, headed by Sir Walter Bodmer, a human geneticist from Oxford and a specialist on the immune system. I presented those aspects of our work that I thought might be appropriate to an anthropology institute, focusing on the study of ancient DNA, especially Neanderthals, and the reconstruction of human history from genetic as well as linguistic relationships between human populations. In addition to my scientific presentation there were several informal discussions about whether anthropology was a topic that the MPS should engage in, given the dire history of the subject in Germany. Perhaps it was easier for me as a non-German born well after the war to have a relaxed attitude toward this. I felt that more than fifty years after the war, Germany could not allow itself to be inhibited in its scientific endeavors by its past crimes. We should neither forget history nor fail to learn from it, but we should also not be afraid to go forward. I think I even said that fifty years after his death, Hitler should not be allowed to dictate what we could or could not do. I stressed that in my opinion any new institute devoted to anthropology should not be a place where one philosophized about human history. It should do empirical science. Scientists who were to work there should collect real hard facts about human history and test their ideas against them.

I did not know how well my arguments went down with the committee. I returned to Munich and months went by, until I had almost forgotten about the whole thing. Then one day I received an invitation to meet with a new MPS committee that had been charged with actually founding an institute devoted to anthropology. There followed a number of meetings with talks by different candidates. The fact that there were no traditions in the subject to build on either in the MPS or indeed in Germany turned out to be somewhat advantageous. It allowed us to discuss freely, unconstrained by academic traditions and preexisting structures, how one would organize a modern institute to study human history. The concept that emerged during our discussions was that of an institute not structured along the lines of academic disciplines but focused on a question: What makes humans unique? It would be an interdisciplinary institute where paleontologists, linguists, primatologists, psychologists, and geneticists would together work on this question. The framework within which one should ask this question was evolution. Ultimately, the goals should be to understand what had set humans on an evolutionary track so different from other primates. So it should be an institute in "evolutionary anthropology."

Neanderthals as the closest extinct relative of modern humans would of course fit well with this concept. So would the study of our closest living relatives, the great apes. And so it came to be that the renowned American psychologist Mike Tomasello, who works with both humans and apes, was invited to start a department in the institute, as was Christophe Boesch, a Swiss primatologist who with his wife Hedwige had spent many years living in the forest in the Ivory Coast to study wild chimpanzees. A comparative linguist, Bernard Comrie, who is British but worked in the United States for several years, was also invited to join the institute. I was very impressed not only by the quality of the people chosen but by the fact that they all came from outside Germany. I, who had lived in Germany for a mere seven years, was the most "German" of the people entrusted with starting this institute. In few European countries could one be so little impeded by chauvinistic prejudice that a huge research institute—which eventually would employ more than four hundred people—would be led entirely by people from outside the country.

During one of the first meetings in Munich at which all prospective directors of departments were present, I suggested that the four of us get out of town to relax and be alone among ourselves. So in the evening we crammed into my small car and drove to Tegernsee in the Bavarian Alps. As the sun was setting we hiked up Hirschberg, a mountain I had often walked and jogged up with friends and students. Most of us were in shoes not at all suited for the endeavor. As the sun set, we realized we would not reach the summit. We paused on a little hill and enjoyed the pristine Alpine landscape. I felt that we were truly connected to one another and that this was a time when people would tend to be truthful. I asked if they were truly going to come to Germany and start the institute, or if they were negotiating with the MPS only in order to try to extort resources from their current institutions in exchange for not leaving, a behavior not uncommon among successful academics. They all said that they would come. Once the sun had disappeared behind the mountaintops, we walked down under tall trees as night was falling. We talked excitedly about the new institute and what we might do there. We all had solidly empirical research programs, we were interested not only in what we did ourselves but in what the others did, and we were all about the same age. I realized that this new institute would happen, and that I would likely be happy there.

There were still many things to work out with the Max Planck Society and among ourselves. A major question was where in the former East Germany the new institute would be located. The MPS had a clear idea. It was to be in Rostock, a small Hanseatic harbor city on the Baltic coast—and the society had a compelling reason. Germany is a federal country composed of sixteen states. Each state pays into the MPS according to the size of its economy. So politicians obviously want as many institutes as possible located in their home states so that they "get their money's worth." The state where Rostock is located, Mecklenburg-Vorpommern, was the only one without a Max Planck Institute, so it obviously had a good reason to demand one. I could sympathize but I felt that our mission was to ensure that the new institute was scientifically successful, not that some political balance among states was upheld. With just about two hundred thousand inhabitants, Rostock was small, had no international airport, and was almost completely unknown outside Germany. I felt it would be hard to attract good people there. I wanted the new institute to be in Berlin. This, I quickly realized, was not going to happen. A huge number of federal institutions had moved there from the former West Germany. To add our institute to the list would be politically impossible and even difficult in practical terms.

The MPS continued to push for Rostock and arranged a visit to the city, where the mayor and his associates would explain the advantages of the place and show us around. I was firmly against Rostock and told the MPS that not only would I not participate in the visit but would be happy to continue to work at the university in Munich. Up to that point, I believed the MPS officials had thought that I was just playing tricks with them when I said I would not move to Rostock. Now, they realized that I would indeed not come if the institute was in Rostock.

Discussion about possible alternative locations followed. To me, two cities in the southern state of Saxony, Leipzig and Dresden, seemed to have good future prospects. They were both fairly large and had a long-standing industrial tradition as well as a state government that was keen on connecting to that tradition. In addition, another Max Planck Institute was being planned for Saxony—one organized by the brilliant Finnish-born cell biologist Kai Simons. I had met him a few times as a graduate student back in the days when I worked on how cells deal with viral proteins, and I was sure that it would become a great institute. My dream was to have the two institutes next to each other to create a campus where synergies between our group and their institute would be possible. Unfortunately, Germany's federal structure thwarted this vision. It was hard enough to argue that the two largest new institutes to be started in East Germany, ours and Kai's, should both be located in the state of Saxony; that they would also be in the same city was totally impossible to imagine. Since Kai and his colleagues were ahead of us in their planning and had already settled on Dresden, we looked at Leipzig. By and large we liked what we saw.

The city had a beautiful city center that had largely survived the war, a great cultural scene with world-class music and art, and, importantly, a zoo that was open to the possibility of cooperating in building a facility where Mike Tomasello could study the cognitive development of the great apes. It also had a large university, which is the second oldest in Germany. During our discussions with the university I came to realize that it had been even more heavily politicized during the time of the German Democratic Republic than other universities, perhaps because it was the center for the sensitive areas of teacher training and journalism studies. Many of the best professors, either by choice or by necessity, had been very involved with the

Communist Party, and after the collapse of the GDR they had been forced to leave their jobs. A few had even committed suicide as a result. The ones who got to keep their jobs were mostly those who'd had bleak academic careers in the GDR. In some cases their careers had been thwarted due to political persecution, but most of them had not distinguished themselves for the same reasons that people in the West do not—lack of talent, lack of ambition, or other priorities in life.

The positions vacated by the politically compromised faculty had mostly been filled by academics from West Germany. Unfortunately, the best people in the west did not tend to make the leap to go east and take on the extra challenges and problems this involved. Instead, those who came were often people who saw this as a chance to get out of a professional dead end. I realized how lucky we were to be able to start an institution from scratch, without the burden of a troubled history. In Dresden the university seemed more ready to take up the challenge of the new times. But we could not have everything. I hoped that over a longer period of time the university in Leipzig would develop the flexibility needed to move forward. On the positive side, Leipzig was a very livable city—even more so than Dresden. I was sure we would be able to convince people to move there. In 1998 our group moved into temporary laboratory space in Leipzig.

We worked hard to get our research going in the new environment and to plan a new large institute building. It was an exhilarating experience. The MPS provided us with ample resources, which enabled me to design a laboratory perfectly suited to our needs and to the ways in which I thought my department should function. This meant, for example, abolishing closed seminar rooms. I decided that the area where our departmental seminars and weekly research meetings were held should be open to the corridor, to do away with the feeling that a meeting was a closed affair for invited participants only. Anyone coming by should be able to listen in, contribute to the discussions, and leave again.

I hoped to attract many people from outside Germany to the institute. I felt that it was very important to create a work environment where scientists and students who had come to Leipzig could develop a social life and a feeling of community with their colleagues and the local students. To help facilitate this I convinced the architects to put in areas in the building for table tennis, table soccer, and even a forty-five-foot climbing wall in the entrance hall. Finally, inspired by the social role of the sauna in my native Scandinavia, I convinced the surprised architects that we needed a sauna on the roof of the building.

But most importantly I could for the first time design a clean room for ancient DNA extractions that was to my specifications. This largely meant giving free reign to my paranoia about contamination from human DNA stuck to dust particles. The "clean room" was in fact not just a single room but several rooms. They would be located in the basement of the building, where you could enter the clean facility without coming even close to laboratories where modern DNA was being handled. In the clean facility, you would first enter a room where you would change to sterile clothing. You would then enter a preliminary room where somewhat "dirty" work, such as grinding bone samples to powder, would be done. From there you would enter the innermost room, where DNA extractions and manipulations of the extracted DNA would be performed. Here, too, the valuable DNA extracts would be stored in special freezers. All work here would be done in hoods where the air was filtered (see Figure 7.1). In addition, the air of the entire facility would be circulated and filtered. It was to be sucked through a grid on the floor, and 99.995 percent of all particulates larger than 0.2 thousandths of a millimeter would be removed from it before it was returned to the room. We constructed not one but two such facilities in the basement so that different types of work—for example, on extinct animals and on Neanderthals—could be separated. No reagents or equipment would ever be allowed to pass from one of the clean rooms to the other, so that if we ever

FIGURE 7.1. The innermost of our clean rooms at the Max Planck institute in Leipzig. Photo: MPI-EVA.

had contamination in one of the rooms the other one would be unaffected. I felt that this facility would finally let me sleep more calmly at night.

Of course, the building and the facilities were of secondary importance to the people who would work there. I looked for group leaders who would work on different but related topics so that the different groups could help and stimulate each other. One scientist I very much wanted to attract to Leipzig was Mark Stoneking. But there were complications.

Mark had done his PhD in Berkeley with Allan Wilson, and it was there that I met him during my postdoc period. He had worked on mitochondrial variation in humans and was one of the main people behind the "Mitochondrial Eve" theory—the realization that the variation in the human mitochondrial genome had its origin in Africa within the last 100,000 or 200,000 years. At that time Mark had worked with Linda Vigilant, a graduate student, using the then-novel PCR to sequence a variable part of the mitochondrial genome from people in Africa, Europe, and Asia. Together with Allan they had published a very influential paper in *Science* that seemed to nail the out-of-Africa story. Although later challenged on statistical grounds, their conclusions nevertheless had stood the test of time. During those heady times in Berkeley, I had been struck by Linda's cute boyish looks when she came into the lab on a motorcycle every day, and by her smarts. But at the time I was emotionally involved both with a boyfriend and with my engagement in the AIDS support group. So I was not crushed when Mark became involved with Linda. They ended up marrying, moving to Penn State University, and having two children. But my connection to Linda was not to end there.

In 1996, six years after I had left Berkeley, Mark, Linda, and their two young boys came to Munich to spend a sabbatical year in my research group. We often made excursions to the Alps together—for example, to my favorite Hirschberg—and they often borrowed my car. Linda did not work in the lab but took care of the kids. In the evenings she sometimes wanted time off from the family and we started going to the cinema together. We got along well and I did not think very much about our relationship until one of my graduate students jokingly said that he thought Linda liked me. This made me aware of the tension between us, most tangibly in the dark movie theaters where we watched alternative European films. One night, in a theater not far from my apartment, our knees touched in the dark, perhaps by chance. Neither of us retracted our knees. Soon we were holding hands. And Linda did not go directly home after the movie.

I had always thought of myself as gay. In the street, I would certainly mostly notice good-looking guys. But I had also been attracted to women, especially those who knew what they wanted and could be assertive. I'd had relationships with two women before. Yet, I thought that being together with Linda, who was married to a colleague and had two children, was not a great idea. It could be a temporary thing at most. But over weeks and months it became more and more clear that we understood each other at many levels, also sexually. Nevertheless, when Mark and Linda returned to Penn State University after their year in Munich, I was sure that my relationship with Linda would end. But that was not to be.

Just as the Max Planck Society started discussing the new institute with me, Penn State contacted me and offered me an attractive endowed professorship. I was torn. I realized that I probably did not want to live in the staid, rural atmosphere of State College. But I also realized that having a serious job offer might make my negotiations with the MPS easier. And another less well-formulated reason may have been at play. I did not mind having to visit State College because Linda was there. I ended up traveling to Penn State a number of times, and Linda and I kept meeting.

This was a difficult time. Not only did I have secrets from Mark, I had secrets with Mark: even as Penn State University was trying to recruit me, I discussed with Mark the possibility of his coming to the new Leipzig institute. All this secrecy and double play finally became too much for me. Perhaps influenced by the double life my father had led (he had two families, one of which did not know about the existence of the other), I had always prided myself on being open and not having secrets in my private life. Yet, here I was living at least a mild version of my father's double life. I convinced Linda that if we intended to continue seeing each other she had better tell Mark what was going on. She did. There was the expected crisis. But the fact that Linda was open with Mark rather early in our relationship may have made the crisis less severe than it would otherwise have been. With time, Mark showed that he was able to separate his private from his professional feelings and after a while he was able to entertain the possibility of moving to Leipzig. Scientifically, this was a great boon for the institute. I was able to convince the MPS to offer him a permanent professorship and create a budget for him. In 1998, when our institute started, Mark, Linda, and their two boys moved to Leipzig and Mark transferred his research group to our institute. Luckily, Linda was also able to find a job at the institute. Christophe Boesch, who was busy planning his department of primatology, was concerned about finding someone who would be able to run a genetics lab focused on wild apes. This would mean relying on strange sources of DNA such as feces and hair left behind by chimpanzees and gorillas in the jungle and collected by the field researchers. Linda had based a large part of her dissertation research at Berkeley on getting DNA out of single hairs for analysis of human genetic variation. I could with a good conscience recommend her to Christophe, and Linda ended up heading up the genetics laboratory in the primatology department.

We all moved into a small apartment building I had bought and renovated. Over the years, Linda and I became closer and Mark found new love while everyday life in our house went on without any great problems. In June 2004, Linda and I were vacationing at Tegernsee. Once again, we were walking down from Hirschberg late one evening. At that point we started talking about the fact that we were getting on in life. We did not have unlimited time in front of ourselves. Unexpectedly, Linda said that if I wanted to have a child she would like one, too. I had played with the idea, and joked about it with her, but now it became clear to me that I very much wanted a child. In May 2005, our son Rune was born.

Over the years that ensued, our lives kept changing, but in small steps. Linda and Mark divorced amicably, and in 2008 Linda and I married. The institute turned out to be a uniquely successful place where researchers, regardless of whether they came from what was traditionally thought of as "humanities" or "sciences," were able to work together. The tradition of hiring the best from all over the world was continued when a fifth department was founded by the French paleontologist Jean-Jacques Hublin. It is a testimony to the attractiveness of our institute that he passed up an almost certain appointment to the College de France, one of the most prestigious institutions in France, to come to the Leipzig institute. In fact, in the fifteen years since the founding of our institute, large universities elsewhere, such as Cambridge University in the UK and Tübingen University in Germany, have copied our concept. Sometimes I wonder why it has worked out so well. One odd reason may be that we were all new to Germany and felt that as we had started this institute together, we had better get along well with each other and make it work. Another may be that even though we are all interested in similar questions, our areas of expertise do not overlap, meaning that there is little direct competition and rivalry among us. Yet another is the generous support from the MPS, allowing us to avoid the petty competition for meager resources that poisons the atmosphere at many universities. Indeed, this has all worked out so well that I sometimes think I should return to the little hill on Hirschberg close to Munich where in 1996 the four founding directors watched the sunset together. I would then erect a small kern as a private little monument to commemorate that something important once happened there. Perhaps one day I will do that.

Chapter 8
Multiregional Controversies

While I was busy planning the new institute and Matthias Krings was trying to retrieve mtDNA from additional Neanderthals, the scientific community had begun to grapple with our analysis of the type specimen from Neander Valley. Our results did not go down well with proponents of the "multiregional-continuity" model of human origins who held, among other positions, that Neanderthals were among the ancestors of present-day Europeans. They shouldn't have been so upset. In our 1997 paper, we'd carefully pointed out that while the Neanderthal mitochondrial DNA was clearly different from the mtDNA of any present-day humans, Neanderthals could nevertheless have contributed other genes—genes in the nuclear genome—to present-day Europeans. Perhaps the multiregionalists' criticism of our work reflected a more general feeling of beleaguerment: as we were showing that, at least for the *mitochondrial* genome, the out-of-Africa and not the regional-continuity model applied, other researchers were finding that patterns of genetic variation in present-day humans supported the out-of-Africa scenario rather than the "multiregional-continuity" scenario. For example, our work was in good company, aligning with the work that Linda Vigilant, Mark Stoneking, and others in Allan Wilson's lab had done in the 1980s on the mitochondrial genome. What's more, we had begun to expand their work to the nuclear genome since I had moved to Germany. And the results seemed clear to me.

This work on the nuclear genome of present-day people was done by Henrik Kaessmann, one of the most talented graduate students I've ever known. Henrik came to the lab in 1997. He was tall, blond, athletic, and extremely serious about his work. I soon took a great liking to running with him in the Alps around Munich, especially up Hirschberg. (This mountain seemed to play a frequent role in my life.) After our strenuous runs up the winding logging roads and our leisurely jogs down them, we would spend time talking about science, especially genetic variation among humans.

We knew, from the work of Allan Wilson and others, that mitochondrial DNA variation was lower in humans than in the great apes, suggesting that humans were special in terms of having expanded from a small population. But we were acutely aware that the small size and simple inheritance of the mtDNA might be giving us a skewed view of the genetic history of humans and apes. By the time Henrik joined our lab, new and faster ways of sequencing DNA had made it possible to study parts of the nuclear genome in present-day people just as we and others had the mitochondrial genome. Henrik wanted to take on this challenge and study nuclear DNA variation in humans and in apes. But which part of the nuclear genome should he focus on?

We understand the function of only some 10 percent of the nuclear genome. These parts mostly contain genes that code for proteins. Such parts of the genome show very few differences between individuals because many mutations are harmful. Also, if a gene changed its function in the past so that the carriers of a new variant survived better or had more children, the gene may have spread in the population and show patterns of differences that reflect this. The rest of the genome is far less constrained by natural selection, presumably because these sequences do not have any essential functions that require DNA sequences to be preserved. Because we were interested in how random variation accumulates over evolutionary time, it was this 90 percent that was of interest to us. We chose to look at a particular region of 10,000 nucleotides on the X chromosome that contained no known genes or other important DNA sequences.

Having determined what part of the genome to sequence, we next turned to the question of which individuals to sequence. Males were the obvious choice because they have only one X chromosome (while females carry two), so Henrik's task would be much simpler. But which males to sequence was a tougher choice. Others had often selected whichever people they had easy access to. For example, many genetic studies (generally of a medical nature) had been conducted using samples from people of European ancestry. A naïve user of databases of human genetic diversity might therefore believe that there is more genetic variation in Europeans than in other groups. But this, of course, may simply reflect the fact that groups other than Europeans had not been studied as much.

We could think of three ways to sample humanity more sensibly. First, we could collect our males based on how many people lived in different parts of the world. This, however, seemed a bad idea since our sample would be predominantly Chinese and Indians, large populations of whom resulted from developments during the past 10,000 years such as the invention of agriculture. In short, we would miss much of the world's genetic diversity. Second, we could collect people according to land area, taking a sample every couple of square miles. But this, in addition to posing formidable logistic challenges, would result in over-sampling of sparsely populated areas such as the Arctic. The third option, which we finally adopted, was to focus on major language groups. We argued that major language groups (such as Indo-European, Finno-Ugric, and so on) reflect some approximation of cultural diversity going back more than 10,000 years. So by focusing on samples representative of major language groups, we could increase our chances of sampling most groups that have had long, independent histories. We would therefore hopefully cover more of human genetic variation.

Fortunately, others had come up with this idea before us so we were able to rely upon DNA samples collected by the distinguished Italian geneticist Luca Cavalli-Sforza at Stanford University. From those samples, Henrik selected sixty-nine men representing all major language groups and sequenced the 10,000 nucleotides in each of them. When he compared the DNA sequences in randomly chosen pairs of men, he found an average of just 3.7 nucleotide differences. Just as had been seen for the mtDNA, he found more variation between pairs of individuals from within Africa than from outside Africa. To gain some perspective on these results, he next turned to the closest living relatives of humans: the chimpanzees.

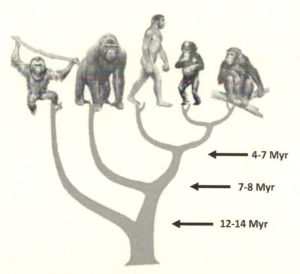

4-7 Myr

7-8 Myr

12-14 Myr

FIGURE 8.1. Tree of humans and great apes indicating approximate times when they may have shared common ancestors (although these dates are very uncertain). Modified from Henrik Kaessmann and Svante Pääbo "The genetical history of humans and the great apes," *Journal of Internal Medicine* 251: 1-18 (2002).

There are two species of chimpanzees, both living in Africa. The "common" chimpanzee lives in equatorial forests and savannahs in a patchy distribution stretching from Tanzania in the east to Guinea in the west, while the bonobo, sometimes called the "pygmy chimpanzee," lives only south of the Congo River, in the Democratic Republic of Congo. Comparisons of DNA sequences had shown that the two chimpanzee species are the closest living relatives of humans, our lineages having split perhaps some 4 million to 7 million years ago. A bit further back, perhaps 7 million to 8 million years ago, humans and chimpanzees shared an ancestor with the other African great ape, the gorilla. Orangutans in Borneo and Sumatra share with the other great apes and humans an ancestor who lived perhaps 12 million to 14 million years ago (see Figure 8.1).

Henrik chose thirty male chimpanzees (the "common" species, not the bonobos), representing the major chimpanzee populations in eastern, central, and western Africa, and sequenced the same stretch of DNA on the X chromosome as he had in the humans. Again making comparisons between randomly chosen pairs, he found an average of 13.4 differences between any two individuals. It was, to my mind, an amazing observation. Seven billion humans hugely outnumber chimpanzees, perhaps numbering fewer than two hundred thousand. And humans live on almost every speck of land there is on the planet while chimpanzees live only in equatorial Africa. Yet any two chimpanzees carried three to four times as many genetic differences from each other than two random humans.

Henrik next sequenced the same piece of DNA in bonobos, gorillas, and orangutans to see whether humans are unusually similar to each other or chimpanzees unusually diverse. He found that gorillas and orangutans carry even more variation than the chimpanzees and that only bonobos had about as little variation as humans. We published these results in three papers in *Nature, Genetics,* and *Science* between 1999 to 2001,[1] demonstrating that a region in the nuclear genome had a pattern of variation very similar to what Allan Wilson's group had found for mtDNA. The same pattern was likely typical of the entire human genome, and I became more convinced than ever that the out-of-Africa model for modern human origins was the correct one. So I listened to the critique of our Neanderthal work from the "multiregionalists" and was not impressed. But mostly I did not answer them. I was convinced that time would tell who was right.

Most multiregionalists were paleontologists and archaeologists. Although I did not dare say so publicly, I privately thought little about their ability to answer questions about whether one ancient group had replaced another one, mixed with it, or simply changed to become the other group. For the most part, paleontologists could not even agree on how to define the ancient groups they studied. There were—and still are—lively fights between "splitters," who see many different species among hominin fossils, and "lumpers," who see few. There are other problems inherent in paleontology. As famously stated by the anthropologist Vincent Sarich, who worked with Allan Wilson in the 1980s, we know that people living today had ancestors because they're here, but when we see a fossil, we cannot know whether it had any descendants. In fact, most fossils we see in museums look like humans because they share ancestors with us sometime in the distant past, but they often have no direct descendants today and represent "dead-end" branches of our family tree. Yet there is often a tendency to think of them as "our ancestors." In my enthusiastic moments, I imagined that the sequencing of DNA extracted from fossils would eventually do away with all this uncertainty.

One of our critics among the multiregionalists was the distinguished paleontologist Erik Trinkaus. He pointed out that our results would be biased if we erroneously discarded as contaminants any DNA sequences resembling those of present-day humans when found in Neanderthal bones. He argued that in fact these may be endogenous, true Neanderthal sequences. Certainly some Neanderthal bones had yielded *only* modern-looking sequences. But these were specimens with bad preservation, so I was confident that all endogenous Neanderthal DNA in them were gone and all we had seen were modern contaminants. Nevertheless, Trinkaus's argument was a logical one and I felt we needed to address it directly.

This became the task of David Serre, a French graduate student from Grenoble with an enormous head of hair and a tendency to ski too fast down mountains in winter and canyon down powerful waterfalls in summer. We decided that his research, should he live to carry it out, would investigate whether all Neanderthals had mitochondrial DNA sequences similar to that from the type specimen, and whether early modern humans in Europe, who lived at the same time or slightly after the Neanderthals, lacked such DNA sequences. The latter question was important to pin down. As noted, the survival of a particular mtDNA sequence in a population is to a large extent influenced by chance. If early modern humans had arrived in Europe and mixed with the resident Neanderthals, then some, or even many, of them might have carried Neanderthal mtDNA sequences that could have become lost in subsequent generations if the females carrying them had no daughters. Indeed, soon after our 1997 *Cell* paper appeared, Magnus Nordborg, a Swedish theoretical biologist working in the United States, had pointed out this scenario.

This criticism did annoy me, because it confused two separate questions. The first question was whether Neanderthals contributed mitochondrial DNA to modern humans that persists in people living today. We had answered this question in the negative. The second question was whether Neanderthals and modern humans interbred. This question we had not answered. However, I found the first question both more interesting and more important. I wanted to know whether I, or anyone else walking around today, carried DNA from Neanderthals in our bodies. If we had *not* inherited any DNA from the Neanderthals, any interbreeding 30,000 years ago was of no consequence from a genetic perspective. Whenever I spoke to journalists, I tried to make this point. To make it clear, I said that I was not in the least interested in sexual practices during the Late Pleistocene unless those practices had left any traces in our genes today. I sometimes added that I would be very surprised if modern humans had not had sex with the Neanderthals they encountered. But what mattered was whether they had children who lived to pass on their genes to us.

Despite my annoyance with these confused questions, I wanted David to investigate whether early modern humans in Europe might have carried Neanderthal mtDNA that subsequently became lost. If they ever did have such mtDNA, they would have carried nuclear DNA from Neanderthals as well. In that case, it would be reasonable to assume that some parts of Neanderthal nuclear DNA might linger on in people today.

We wrote to several museums all over Europe to collect Neanderthal and early modern human bones. After our success with the Neanderthal type specimen it had become somewhat easier to convince curators to allow us to sample their collections, and we ended up with bits of bones from twenty-four Neanderthals and forty early modern humans. David analyzed the amino acids in all sixty-four samples. Just four Neanderthal and five early modern human samples were well enough preserved to suggest the presence of mtDNA—a grim but typical proportion. He extracted DNA from these nine bones and tried to do PCRs using primers that could amplify mitochondrial DNA from the great apes, Neanderthals, as well as humans. David got amplification products from all nine samples. When sequenced, they turned out to be similar or identical to those found in people today. These results were disturbing. Maybe Trinkaus was right after all.

I asked David to do the experiment again, this time including samples from five cave bears from Vindija and one from Austria. When he amplified these, they also yielded human sequences! This strengthened my suspicion that we were simply obtaining contaminating DNA sequences from modern humans who had handled these bones. David then carefully designed primers that would amplify Neanderthal-like mtDNA but not present-day human mtDNA. After using mixtures of DNAs in the lab to test that these primers were indeed specific for Neanderthal mtDNA, he tried them on the cave bears. He could amplify nothing. This was reassuring. The primers were indeed specific for Neanderthal mtDNA. He then used these primers on the extracts from the Neanderthal and modern human bones. Now all the Neanderthal bones yielded mtDNA sequences similar to that of the Neanderthal type specimen, suggesting once again that Neanderthals did not carry mtDNAs similar to those of present-day humans. In contrast, none of the five early modern humans yielded any products, suggesting that Trinkaus was wrong.

We next turned to theory to explore this topic further. We designed a population model in which we assumed that Neanderthals bred with anatomically modern humans 30,000 years ago and that those modern humans had descendants living today. We then asked what the biggest genetic contribution to present-day humans could be, given our findings that neither any humans today nor five early modern humans some 30,000 years ago carried any Neanderthal mtDNA. According to this model (which we made tractable by using simplifying assumptions, such as not incorporating modern human population growth), Neanderthals could have contributed no more than 25 percent to the nuclear genome of people living today. However, because we saw no direct evidence for a genetic contribution from Neanderthals, I felt that the most reasonable hypothesis, unless new data showed something different, was that Neanderthals had made no genetic contribution to people alive today.

I found that this result nicely illustrated the strength of our approach as compared to a typical paleontological analysis. We used clearly defined assumptions and drew conclusions that were bounded by probabilities. Nothing approaching this in rigor could be done using morphological features of bones. Many paleontologists liked to portray what they did as rigorous science, but the very fact that they had been unable to agree on the occurrence of a genetic contribution from Neanderthals to present-day humans despite at least two decades of debate illustrated that their approach had big limitations.

After we published David's results,[2] a theoretical group in Switzerland, led by the population geneticist Laurent Excoffier, developed a much more plausible model than ours for how Neanderthals and modern humans might have interacted. They assumed that when anatomically modern humans moved across Europe, any interbreeding with Neanderthals would have taken place in the areas at the front of the modern human advance. This initial invasion would have been characterized by small modern human populations that would then increase in size. The Swiss group showed that under this model even rare instances of interbreeding would have been likely to leave traces in today's mitochondrial gene pool, because on average females in a growing population have multiple daughters who will transmit their mother's mtDNA. So under such circumstances, any Neanderthal mtDNA that had ended up in the modern human population would run much less risk of getting lost than if this population were of constant size. Since we had seen no Neanderthal mtDNA in either the five early modern humans or the thousands of living humans we and others had studied, Excoffier's group concluded that our data suggested "an almost complete sterility between Neanderthal females and modern human males, implying that the two populations were probably distinct biological species."[3]

I had nothing against this conclusion from the Swiss group, but it was of course still possible that something special, not captured by their model, had gone on when Neanderthals and humans met. For example, if all children of mixed Neanderthal-modern human ancestry ended up in Neanderthal communities, they would not have contributed to our gene pool and the result would look like "almost complete sterility" as that group described it. Also, if all interbreeding events involved Neanderthal males and modern human females, they would not be detectable in today's mtDNA gene pool, since males do not contribute mitochondrial DNA to their children. Such mixtures would be seen only in the nuclear genome. To fully understand how interactions between our ancestors and the Neanderthals may have impacted our genomes, we clearly needed to study the Neanderthal nuclear genome.

Chapter 9
Nuclear Tests

Henrik's work on the X chromosome had shown that the patterns of similarities and differences seen in the mitochondrial DNA of humans and apes were extendable to at least one part of the nuclear genome. Whether we would ever be able to study nuclear DNA from Neanderthals or be forever limited to their mitochondrial genome was not at all apparent. In my darker moments, I thought we were going to be stuck with mtDNA's blurry, one-eyed view of human history. Certainly, if one disregarded results from animals and plants embedded in amber, dinosaurs, and other fanciful "antediluvian" studies (which I did), nobody had yet succeeded in retrieving any nuclear DNA from ancient remains. But in my more considered moments, I felt that we should give it a try.

It was at this point that Alex Greenwood, a diminutive but determined new postdoc from the United States, arrived in the lab. I told him about our hopes for retrieving nuclear DNA from Neanderthals, noting that it was a high-risk project but also a very important one. He was eager to take on the challenge.

I suggested a "brute-force" approach. My plan was to test samples of many bones to find those with the most mtDNA and then extract DNA from yet larger samples in an attempt to retrieve any nuclear DNA. This approach meant that we could not perform our initial experiments with the uncertain technique on Neanderthal remains; they were too scarce and valuable to use when the risk of failure was so high. Instead we resorted to animal bones, which were both considerably more abundant and less valuable to paleontologists. The cave-bear bones I had brought back from the dark basement of the Quaternary Institute in Zagreb now came in handy. They had been found in Vindija Cave, a limestone cave that had also produced some Neanderthal remains that contained mtDNA. So if we were able to retrieve nuclear DNA from the cave bears, we could hope to do the same with the Neanderthals.

Alex began by extracting DNA from the Croatian cave-bear bones, which were between 30,000 and 40,000 years old, and checked to see if they contained any bear-like mtDNA. Many of them did. He then took the extracts that seemed to contain the most mtDNA and tried to amplify short fragments of nuclear DNA. This failed. He was frustrated, and I was dismayed but not surprised. The problem he faced was a familiar one to me: because each cell in a living animal contains hundreds of mitochondrial genomes but only two nuclear genomes, any particular piece of nuclear DNA was present in 100- or 1,000-fold fewer copies in the extracts than any particular piece of mitochondrial DNA. So even if some nuclear DNA was present in minute amounts, the chances of amplifying it were a 100- or 1,000-fold lower.

One obvious way to overcome this problem was to simply use more bone. Alex made extracts of ever larger amounts of cave-bear bone and tried amplifying ever shorter pieces of nuclear DNA using primers flanking nucleotides where bears were known to differ from humans. That would enable him to discriminate between ancient bear DNA and contaminating human DNA. But in these mega-extracts, *nothing* could be amplified—not even bear mitochondrial DNA. He got no products at all.

After several weeks of repeated failures with multiple bones, we realized it was impossible to make useful DNA extracts from such large amounts of bone material. This was not because the bones contained nothing to amplify but because the extracts contained something that inhibited the enzyme used for the PCR; it became inactive and no amplification at all took place. We struggled to purify the unknown inhibitor away from the DNA in the extract but failed. We diluted the extracts in small steps until they started working again for the amplification of mtDNA. Then, at that dilution, we tried the nuclear amplification. It always failed. I tried to remain upbeat but as the months passed, Alex became more and more frustrated and anxious about whether he would ever produce any results that would justify a paper. We began to wonder if after a bear's death the nuclear DNA might be degraded by enzymes leaking through the nuclear membrane of the decaying cells. Perhaps the DNA in mitochondria, having a double membrane, would have been better protected, making the mtDNA more likely to survive until the tissue dried out, froze, or was otherwise protected from enzymatic attack. This possibility made me wonder whether it would be possible to find nuclear DNA in ancient bones at all, even if we could overcome the inhibition of the PCR. I was slowly becoming as frustrated as Alex.

Thwarted by the cave bears, and wondering whether the conditions in the cave may simply have been too unfavorable to preserve nuclear DNA, we decided to switch to material that we expected to show the very best preservation—permafrost remains of mammoths from Siberia and Alaska. These had been frozen ever since the animals died and freezing, of course, will slow down and even stop both bacterial growth and many chemical reactions, including those that degrade DNA over time. We also knew, from Matthias Höss's work, that mammoths from the Siberian permafrost tended to contain large amounts of mtDNA. Of course, no Neanderthals had ever been found in the permafrost—so switching to mammoths meant taking a step away from my ultimate goal. But we needed to know whether nuclear DNA could survive over tens of thousands of years. If we found no nuclear DNA in the frozen remains of mammoths, then we could forget about finding it in Neanderthal bones preserved under much less ideal conditions.

Fortunately I had systematically collected ancient bones from different museums over the past few years, so Alex could immediately try remains of several mammoths. He found one mammoth tooth that contained particularly large amounts of mtDNA. It had been pulled out of the frozen ground when the Alaska Highway, extending from northeastern British Columbia to near Fairbanks, was built in great haste during World War II and stored ever since in a huge box in the American Museum of Natural History. To make the search for DNA a bit easier, we carefully targeted a segment of the nuclear genome containing part of the gene known as 28S rDNA, which encodes an RNA molecule that is part of the ribosome, a structure that synthesizes proteins in cells. For our purposes, this gene had the great advantage of existing in a few hundred copies per cell. It should thus have been about as abundant as mitochondrial DNA in the extracts, assuming that the nuclear DNA had not been degraded more than mitochondrial DNA after death. To my delight and profound relief, Alex could amplify the ribosomal gene. He sequenced clones of the mammoth PCR products and reconstructed the gene's sequence using the overlapping-segments approach we had established when studying Neanderthal mtDNA. He then wanted to compare the sequence to those from African and Asian elephants, the closest living relatives of mammoths. I had been so paranoid about contamination that, until Alex had the mammoth results, I had forbidden him or anyone else to work on elephants. But now, working outside our clean room, Alex used the same primers he had used for the mammoth work to amplify and sequence the 28S rDNA fragment from an African and an Asian elephant. The mammoth sequences were identical to those of the Asian elephant but differed at two positions from the African elephant version, suggesting that mammoths were more closely related to Asian than to African elephants. But comparing mammoths to living elephants hadn't been the point of the exercise: finding ancient nuclear DNA had been. To clinch it, we sent a bit of the tooth off for carbon dating. When the 14,000-year-old date came back, I felt satisfied for the first time in months. It was now official. These were the first nuclear DNA sequences ever determined from the late Pleistocene.

Encouraged by these results, Alex designed primers for amplifying two short pieces of a fragment of the von Willebrand factor gene, only one copy of which exists in the elephant genome. This gene, abbreviated vWF, encodes a blood protein that helps platelets stick to wounded blood vessels. We focused on it because others had already sequenced it from elephants (and many other extant mammals), so if we managed to determine a sequence from the mammoth, we could directly compare it to those present-day sequences. I could hardly believe my eyes when, during our weekly lab meeting, Alex showed pictures of bands in a gel that suggested he was able to amplify these gene fragments from the mammoth. He repeated the experiment twice, using an independently prepared extract from the same mammoth bone. Among the many clones he sequenced, he saw errors in individual molecules, presumably due either to chemical damage in the old DNA or to the DNA polymerase's addition of the incorrect nucleotide during the PCR cycles (see Figure 9.1). At one position, however, he saw an interesting pattern. He had sequenced a total of thirty clones from three independent PCR amplifications. At one position, fifteen of those clones carried a C, fourteen carried a T, and one carried an A. The single A, we assumed, was an error caused by the DNA polymerase, but the others represented something that made my heart beat a bit faster. This particular spot in the sequence was clearly what geneticists call a heterozygous position, or a single nucleotide polymorphism (SNP for short), a place where the two copies of the gene that this mammoth had received from its mama and papa mammoth differed. What we saw was the first heterozygous position or SNP from the Ice Age. This was, if you like, the essence of genetics—a nuclear gene that has two variants in a population. Things were looking up. If we could see both versions of this mammoth gene, then all parts of the genome must be potentially accessible. It should thus be possible, at least in theory, to retrieve any genetic information we wanted from a species that went extinct many thousands of years ago. To drive home this point, Alex

```
Mammoth,                                         ↓
consensus sequence
allele 1:          .....-....-.............G.A.........................C.
allele 2:          .....-....-..T..........G.A.........................C.

Mammoth,           .....-....-.............G.A.A...............A......C.
clones:1st extract, .....-....-.........T...G.A.........................C.
1st PCR            .....-....-..T..........G.A.........................C.
                   .....-....,-......N......G.A.A...............A......C.
                   .....-..T-..............G.A.AA..............A......C.
                   .....-....-..T..........G.A.....A...................C.
                   .....-....-.............G.A.A...............A......C.
                   .....-....-.............G.A.A...............A......C.
                   .....-....-.............G.A.A...............A......C.

Mammoth,           .....-....-..T.......T...G.A...............G........C.
clones:2nd extract, .....-....-..T........TG.A.........................C.
1st PCR            .....C...TN..T.....T.T..G.A.........................C.
                   .....-....-.............G.A..........C.............C.
                   .....-....-..T..........A.A.........................C.
                   .....-....-.............G.A.........................C.
                   .....-....-..T..........G.A.........................C.
                   .....-....-..A..........G.A....A....................C.
                   ..N..-....-..T..........G.A.........................C.

Mammoth,           ...T.-....-.............G.A.........................C.
clones:2nd extract, .....-....-.............G.A.........................C.
2nd PCR            .....-....-..T..........G.A.................T.T..C.
                   .....-....-.............G.A.A.AA...................C.
                   .....-....-..T..........G.A.........................C.
                   .....-....-..T...N......G.A.N................T......C.
                   .....-....-..T..........A.A.................A......C.
                   .....-....-..T..........G.A.............T.......T.T..T.
                   .N...C..-..-.............G.A.........................C.
                   .....-....-.............G.A.........G...............C.
                   .....-....-......T,.T..TG.A.................T......C.
                   .....-....-..T...TT....TG.A.........................C.
```

FIGURE 9.1. Cloned DNA sequences from three amplifications of a nuclear gene fragment from a 14,000-year-old mammoth. The arrow points to the first heterozygous position or SNP ever observed from the Late Pleistocene. From A. D. Greenwood et al., "Nuclear DNA sequences from Late Pleistocene megafauna," *Molecular Biology and Evolution* 16, 1466–1473 (1999).

amplified pieces of two more single-copy genes: one encoding a protein regulating the release of neurotransmitters in the brain and one encoding a protein that binds vitamin A and is produced by the rods and cones in the eyes. He was successful in both cases.

Since we had struggled so long to retrieve nuclear DNA, Alex's mammoth results were very welcome indeed and for several days I was very happy about them. But of course I wasn't all that interested in mammoths. I was interested in Neanderthals, and I was painfully aware that there were no Neanderthals in permafrost. I urged Alex to go back and try the cave-bear remains from Vindija again, to see if he could retrieve nuclear DNA from remains that were not frozen. He analyzed mitochondrial DNA from several Croatian cave bears and identified one bone that seemed to contain a lot of it. We carbon-dated it and found it to be 33,000 years old, and thus roughly contemporaneous with the Neanderthals. Alex concentrated on this bone. He tried the ribosomal RNA gene that occurs in many copies in the genome. He obtained small amounts of an amplification product. Reconstructing its sequence from clones, he found that the sequence was indeed identical to that in present-day bears.

This was a success, but one with a dark side. It was so hard to amplify this multicopy-gene fragment that any attempt to get single-copy genes such as the vWF gene he studied in the mammoth seemed doomed to fail. Alex tried it anyway, of course, but as expected he was unsuccessful. So, after all the excitement over the mammoth results, I was—secretly—deeply disappointed by these experiments. We had demonstrated that nuclear DNA could survive over tens of thousands of years in permafrost, but only traces of very common nuclear DNA sequences could be found in the bones of cave bears. There was an enormous difference between permafrost and limestone caves.

In 1999, we published Alex's findings in what I considered to be a beautiful paper, though it was subsequently largely overlooked.[1] It demonstrated that nuclear DNA survives in remains found in permafrost and that even heterozygous positions, where the two chromosomes in an individual differ in DNA sequence, can be determined. We were optimistic about the prospects for genetic research in the permafrost, and we noted at the end of the paper that

> A plethora of faunal remains exist in permafrost deposits and other cold environments. The fact that such remains can yield not only mtDNA but also single-copy nuclear DNA sequences in a substantial amount of cases opens up the possibilities of using nuclear loci in phylogenetic and population genetic studies and of studying genes determining phenotypic traits.

Eventually others would take up this line of work, although not for another five to ten years. Worse, barring the discovery of a Neanderthal in permafrost, it seemed we might never see the whole genome of a Neanderthal.

Chapter 10
Going Nuclear

In the lab, I went about the business of supervising the experiments that pushed the work slowly but reliably forward. But whenever I was confined to a small seat during a long airplane ride, or to a darkened lecture hall during a seemingly irrelevant presentation at a conference, I came back to my greatest frustration: our inability to retrieve nuclear DNA from Neanderthals. I felt that it *had* to be there, even if the PCR could not retrieve it. We simply had to come up with a better way to find it.

A fresh attempt in this direction was made by Hendrik Poinar. Worn down by his fruitless quest for DNA from fauna and flora encased millions of years ago in amber, he had decided to move on to more promising endeavors. Fortunately for us I had just spent time at some boring conference lectures, where I got to thinking about the work we'd done on retrieving DNA from animal droppings. One of those we'd studied was the extinct American ground sloth, an Ice Age animal. The giant sloths had left behind large amounts of droppings, which archaeologists dressed up with the fancy name of coprolites. In fact, in some caves in places like Nevada, the entire floor, to some depth, is largely made up of old ground-sloth feces. In a paper in *Science* in 1998, Hendrik had already shown that mitochondrial DNA was preserved in such material and we described the retrieval of plant DNA from a single sloth bolus, showing how it could be used to reconstruct the ingredients of the meal the sloth had ingested shortly before its death 20,000 years ago.[1] This success suggested that lots of DNA, even nuclear DNA, was preserved in ancient fecal remains. I suggested that Hendrik try to find it.

Hendrik had begun the search with a chemical trick we had developed the year before. Back in 1985, when I was analyzing the mummies from Berlin, I had noticed that almost all of the extracts contained a component

that produced blue fluorescence in UV light, and that if an extract fluoresced blue, it never yielded DNA. I did not know what this component
was, but the observation was painfully memorable because of the disap-
pointment I felt when I saw the blue instead of the pink glow I was hoping
for. As I learned more about the chemistry that may have gone on in dead
tissues over thousands of years, I came across the phenomenon known as
the Maillard reaction, a chemical reaction much studied by the food industry. As it happened, my mother was a food chemist, so she sent me lots
of literature on it. The Maillard reaction occurs when common forms of
sugar are heated or persist at less-hot temperatures for a long time. They
then form chemical cross-links with amino groups found in proteins and
in DNA, resulting in large, tangled molecular complexes. The Maillard reaction occurs in many forms of cooking, and side products of the Maillard
reaction result in the pleasant smell and color of freshly baked bread. But
most interesting to me was that Maillard products give off a blue fluorescence in UV light. I thought that this might have been what was going on
with the Egyptian mummies. I associated this reaction not only with the
blue fluorescence of the mummy extracts but also (perhaps incorrectly)
with their brown color and their characteristic smell, which is sweet and
not unpleasant. And I wondered if the reason I couldn't extract DNA from
them was that the Maillard reaction bonded it to other molecules.

There was a way to find out. In 1996, a paper in *Nature* described a
chemical reagent—N-phenacylthiazolium bromide, abbreviated PTB—
that could break down the complexes formed by the Maillard reaction.[2]
PTB, when added to baked bread, would turn it into dough again (albeit,
surely, not dough that anyone would be tempted to put back in the oven).
Since PTB could not be bought commercially, Hendrik synthesized it in
the lab. When we added PTB to extracts from ancient samples of cave
bears and Neanderthals, it indeed sometimes resulted in better amplification. And when Hendrik added PTB to extracts from the 20,000-year-old
Nevada coprolites, he was able to amplify fragments of the vWF gene that
Alex had partially sequenced from mammoths as well as fragments from
two other nuclear genes, all to my great surprise. We published this work
in July 2003,[3] finally demonstrating that the nuclear genome could be preserved even when remains were not frozen.

Encouraged by those results, I felt that there was now every reason
to persist in our attempts to retrieve nuclear DNA from the bones of cave
bears—again, with PTB. But sadly, this time the chemical trick didn't help.
Actually, it turned out that the Nevada coprolites were a rare exception

where PTB could turn failure into success. The coprolites did, however, confirm my feeling that the nuclear DNA was there, and that we simply needed new techniques to find it.

To get ideas about new techniques I took to consulting as many people as possible about ways to sequence small amounts of DNA. One of the people subjected to my questioning was the Swedish biochemist Mathias Uhlén, a creative inventor and biotech entrepreneur. Mathias combines an apparently unlimited energy and a childlike enthusiasm for new ideas with a knack for collecting creative people around him and transmitting his enthusiasm to them. I always felt energized after my encounters with him. One of the many creative people around Mathias was Pål Nyrén. Ten years earlier, he had conceived of a new technique for DNA sequencing he had developed in spite of widespread skepticism. Mathias had realized the potential of Pål's idea. He also saw that it was timely to think about new ways to sequence DNA: we were still using the approach invented by Fred Sanger in the UK, which, in 1980, had earned him his second Nobel Prize in chemistry.

The Sanger sequencing method relies on the sequential incorporation of the four nucleotides by a DNA polymerase, an enzyme that makes new DNA strands using old ones as templates. In such a sequencing reaction, the DNA polymerase starts its synthesis of DNA strands from a primer at a defined point in the DNA. A small fraction of each of the four nucleotides are labeled with different fluorescent dyes and chemically modified so that when the DNA polymerase builds them in, the synthesis stops. This process creates DNA strands of different lengths, each with the dye at its end indicating which nucleotide sits there. The fragments thus terminated and labeled can be separated by electrophoresis in a gel according to their size. This reveals which dye and thus nucleotide is present—for example, ten positions away from where the synthesis started, eleven positions away, twelve positions away, and so on. The best machines employed to sequence DNA—for example, by the Human Genome Project—can sequence almost a hundred pieces of DNA at a time, for stretches as long as 800 nucleotides. What Pål developed in Mathias's lab was a method called pyrosequencing. Though still in its infancy, this method was potentially much faster and simpler than Sanger sequencing.

Pyrosequencing also uses a DNA polymerase to build DNA sequences, but it detects each nucleotide incorporated into the DNA not by cumbersome separation of fragments according to size but by a flash of light emitted after each nucleotide is built into the DNA chain. The trick Pål had devised was to add just one of the four nucleotides at a time to the reaction mix. For example, if an A (adenine) is added, and the strand that is being used as a template at that point carries a T (thymine, which pairs with adenine), the DNA polymerase builds the A into the growing strand, and an enzymatic system in the reaction causes a light signal to be generated. This flash of light is detected by a powerful camera and registered by a computer. If the template strand carries not a T but another base, no flash of light is generated. Pål added each of the four nucleotides consecutively, cycle after cycle. By noting the light flashes, he could read the order of nucleotides in a DNA fragment. It was a brilliant method, relying only on pumping the nucleotides and other reagents into a reaction chamber and taking pictures with a camera. Even more importantly, it could easily be automated. When Mathias told me about it, I became as enthusiastic as he was.

A little later, Mathias asked me to serve on the scientific advisory board of a company called Pyrosequencing, which he and Pål had founded to produce a commercial instrument for performing this technique. I gladly agreed, since doing so would give me an opportunity to keep up with the development of an exciting technology that I thought might transform how we studied ancient DNA. I joined the advisory board in 2000, a year after the company had produced its first commercial instrument, which could simultaneously sequence ninety-six different DNA fragments, each isolated in a well on a plastic plate. However, from each fragment it could read only about thirty consecutive nucleotides. This was deeply unimpressive compared with contemporary machines relying on the Sanger principle, but pyrosequencing was a young technology and had not reached the limits of its possibilities. In fact, although I did not fully appreciate this at the time, it represented the beginning of a revolution, known as "second-generation sequencing," that would fundamentally change not just our investigations of ancient DNA but many aspects of biology.

I very much wanted to try out pyrosequencing, so I asked Henrik Kaessmann to spend some time in Mathias's lab at the Royal Institute of Technology in Stockholm. Henrik welcomed the opportunity to surprise people in Stockholm with his flawless Swedish; although he grew up in southern Germany, he speaks the language fluently thanks to his Swedish mother. He was also able to generate data from present-day human populations in Europe and Asia that would help in showing how they were related to one another. As with all new techniques, this required learning new skills and some troubleshooting, but it worked well.

In August 2003, the board of Pyrosequencing decided to license the technology to 454 Life Sciences, a US company founded by the biotech entrepreneur Jonathan Rothberg. 454 Life Sciences intended to enhance pyrosequencing with state-of-the-art fluidics. Its innovation relied on adding short synthetic pieces of DNA to the ends of DNA molecules. Single strands of DNA were then captured on beads and ingeniously amplified in little oil bubbles, allowing hundreds of thousands of different strands to be amplified separately but simultaneously in one big reaction. Then the beads were separated from one another on a plate with hundreds and thousands of wells for the pyrosequencing step. Finally (and crucially), to keep tabs on which wells were emitting flashes of light from cycle to cycle, the company used image-tracking methods borrowed from astronomers, who track millions of stars in the night sky. This allowed it to simultaneously sequence not ninety-six but two hundred thousand DNA fragments at a time!

Given that kind of power, I thought perhaps we could simply sequence the random DNA fragments we had in an extract from an ancient bone and see everything that was in there. This brute-force approach would be completely different from the PCR-based method, where we tried to fish out each piece of sequence we wanted to study. The PCR method was not only tedious but also (since we had to decide in advance what to look for) effectively blinded us to all other sequences in the extract. Although the 454 Life Sciences instruments could not sequence DNA fragments longer than 100 nucleotides, the nuclear DNA fragments we had seen in Alex's work on the mammoths and Hendrik's work on the ground sloth had never been longer than 100 nucleotides anyway. I longed to try out a 454 machine.

Mathias and the people working on pyrosequencing were not the only ones I talked to about new approaches. Another was Edward M. Rubin, a dynamic and ebullient genomicist who visited our lab in Leipzig in July 2005. I was eager for his advice. A professor at Lawrence Berkeley National Laboratory in Berkeley, California, and director of the US Department of Energy's Joint Genome Institute, Eddy was certain that the way forward was to clone the DNA in bacteria, by much the same methods as I had used back in the 1980s when I was working with mummies in Uppsala. These methods, he told me, were now much more efficient than they had been back then. I agreed to try this out on cave bears, and we made extracts from two fossilized cave-bear bones we knew contained lots of bear mtDNA and sent them to Eddy's lab in Berkeley. There the DNA molecules in the extracts were fused to carrier molecules, just as I had done back in 1984, and introduced into bacteria. When the bacteria grew, they constituted what is called a "library," in which each bacterial colony, or "clone," contained millions of copies of one unique DNA molecule from the cave-bear bone extract. The DNA from each such colony in the library could be isolated and sequenced, and thus "read," just like a book in a library. Eddy's people used traditional Sanger chemistry to sequence about 14,000 random DNA clones from two such libraries—orders of magnitude more than had been possible back in 1984. From the 14,000 clones, a total of 389, only 2.7 percent, carried DNA sequences that were similar to those found in dog DNA and, therefore, were likely to have come from the cave bear. The rest were from bacteria and fungi that had colonized the bones after the animal's death. Although the proportion of endogenous DNA in these bone extracts was pathetically small, the result was nonetheless exciting because it showed that bones from European caves did indeed contain some nuclear DNA.

We published this result, with Eddy and his Berkeley group as the main authors, in *Science* in 2005.[4] In that paper we somewhat grandiosely claimed that this meant that genome sequencing from ancient remains was possible. But after the paper was published, some people in my own group considered more deeply what had been done, did some calculations, and came to a sobering conclusion. The Berkeley group had sequenced every bit of DNA libraries we had sent, and found a total of 26,861 nucleotides from the cave-bear genome. Given that we had used a few tenths of grams of bones for making these libraries and that the genome is composed of some 3 billion nucleotides, we would have to use more than a hundred thousand times more bone than we had already—more than ten kilos, or twenty-five pounds of bone, in other words—to arrive at even a rough overview of the cave-bear genome. Grinding up that much bone and transforming it into extracts for sequencing libraries would be feasible, if tedious in the extreme, but the massive amount of sequencing then required would be very expensive. And even if it worked, barring unforeseen technical breakthroughs it would never be possible for us to apply this brute-force approach to the truly interesting fossils for which we had only minuscule samples. Sequencing a Neanderthal genome by cloning in bacteria did not seem the way to go, at least to me. Indeed, it seemed impossible. I imagined that most of the DNA must be lost when the bacterial libraries were constructed, probably because it never entered the bacteria in the first place or because it was chewed up by enzymes inside the bacteria. Eddy, however, continued to be enthusiastic about it and suggested that the low efficiency with which the DNA sequences were produced from the DNA extracts was unusual. He argued that future tries would certainly work better and require less starting material.

Despite Eddy's enthusiasm, and in addition because I was averse to relying upon only one approach, I was certain that we needed to try pyrosequencing. It seemed feasible to directly apply the 454 version of pyrosequencing to all the DNA in an extract, thus eliminating the losses caused by getting the DNA into temperamental bacteria. What's more, Jonathan Rothberg and 454 had produced a machine that could sequence hundreds of thousands of DNA molecules in a day. But it was not easy to contact him, as he wisely eschewed easy contact to insulate himself from the crank scientists who might otherwise barrage him with demands for access to his new technology. I tried several avenues and got nowhere. Finally I talked to Gene Myers, the bioinformatics wizard who had helped the famous genomicist Craig Venter assemble the human genome in 2000. I had met Gene at a bioinformatics meeting in Brazil in 2001 and immediately liked his irreverent attitude toward any problem with which he was confronted. We had bonded over a shared interest in skiing and scuba diving. Gene was now a professor at UC Berkeley and an adviser to Rothberg's company, so in July 2005 he was able to put me in e-mail contact with Jonathan.

Jonathan arranged a conference call with me and Michael Egholm, a Danish scientist who ran the operations at 454. When Jonathan got on the line, I began to worry. He was as energetic and intense as I had expected from an entrepreneur of his caliber, but he seemed interested in only one thing—sequencing dinosaur DNA! I was unsure how to handle this annoying predilection, since I was on record as saying that sequencing dinosaur DNA was and would remain impossible. I tried to reiterate that assertion without burning my bridges, emphasizing that there were other cool genomes one could sequence, particularly that of Neanderthals. Fortunately, Jonathan quickly became intrigued by the idea that we might use such an investigation to identify what changes made us fully human. I also convinced him and Egholm that it would be good to start with a mammoth and a cave bear.

A week later we shipped off a mammoth extract and a cave-bear extract to 454 Life Sciences. At about the same time, Richard E. (Ed) Green, a hard-working and talented bioinformatician, joined our lab from UC

Berkeley, where he had just finished his PhD. He had been awarded a prestigious, well-paid fellowship from the National Science Foundation in the United States to undertake a project comparing RNA splicing in humans and great apes. Splicing is the process by which the RNA copies of genes are cut up and joined to form the messenger-RNA molecules that direct protein synthesis. The idea was that differences in how genes were spliced together might account for many of the differences between humans and chimpanzees. But just as Ed got this under way, the first data from 454 Life Sciences came in.

The people at 454 had produced DNA sequences from hundreds of thousands of pieces of DNA from the mammoth and cave-bear bones. I asked Ed to look into the first problem with the DNA sequences, separating those that came from the specimens themselves and those that came from contaminating bacteria and other organisms. It was not a trivial issue. He compared the DNA sequences from the bones to the genome sequences of the elephant and the dog, the two present-day animals most closely related to mammoths and cave bears, respectively, with available genome sequences. But the ancient DNA sequences were short and they were likely to carry errors induced by chemical modifications that had occurred over millennia. In addition, the number and identity of the bacteria and fungi in the bones were unknown. But the challenge of this ancient DNA sideline proved irresistible to Ed; soon he had forgotten all about RNA splicing. Eventually he wrote a letter to the administrator who handled his fellowship at the National Science Foundation, describing how the goals of his project had shifted. Unfortunately, the NSF lacked the vision to recognize that the Neanderthal genome was an awesome opportunity for a computational biologist; instead, it cut his fellowship. Fortunately, our budget was large enough to allow us to keep Ed.

He had in the meantime found that about 2.9 percent of the DNA extracted from the mammoth bone had actually come from the mammoth and about 3.1 percent of the DNA from the cave-bear bone had come from the cave bear. This meant that our earlier results when working with Eddy Rubin, in which only about 5 percent of the cave-bear sequences produced after cloning actually came from the cave bear, were actually pretty good. Three or 5 percent doesn't sound like much, but in total we now had 73,172 different mammoth DNA sequences and 61,667 different cave-bear sequences. This meant that in one single experiment, for which we did not even use up the entire extract, the 454 approach had produced almost ten times as much data as we had obtained from the bacterial cloning of the

DNA from the cave bear in Berkeley. This seemed like a real breakthrough to me, but the approach was not without risk. Our original, PCR-based method enabled us to repeat the experiments many times, both to make sure we got the same sequences and to detect errors in them. Our new approach let us see each sequence only once, and because both genomes were so big, we were unlikely to see another copy of the same segment in the mammoth or cave-bear genome among our sequences. As a result, we couldn't immediately determine the extent to which chemical damage in the ancient DNA, and resultant errors in the sequence, might influence our results.

Detecting errors was not a new problem, however, and we had already made some progress. A few years earlier, in 2001, Michael Hofreiter, then a graduate student in my lab, had together with others in the group shown that the most common form of DNA damage resulting in errors in ancient DNA sequences was the loss of an amino group from the nucleotide cytosine. This occurs spontaneously in DNA whenever even small amounts of water are present. When cytosine (C) loses its amino group, it becomes uracil, a nucleotide usually found in RNA. DNA polymerases read it as a T. By comparing our mammoth and cave-bear sequences to those from the elephant and dog, we could check whether we were seeing more T's than expected in places where the present-day animals had C's. We did see a clear overabundance of T's. But to our surprise we saw also a more modest increase in guanine (G) relative to adenine (A), suggesting that A's might also lose their amino groups in ancient DNA, just as C's did. To test this notion, we used synthetic pieces of DNA into which we introduced both C's and A's without their respective amino groups, to see how they would be read by the DNA polymerase that 454 Life Sciences had used to amplify the DNA in pyrosequencing. The DNA polymerase not only read C's without amino groups as T's but also A's without amino groups as G's. So we wrote in a paper that appeared in the *Proceedings of the National Academy of Sciences* in September 2006 that not just C's but also A's might lose their amino groups.[5] Rather soon, however, it would turn out that we were wrong.

In the meantime, subtle frictions had developed between my group and Eddy Rubin's group at Berkeley. It was now clear to us in Leipzig that pyrosequencing was at least ten times more efficient than bacterial cloning. It seemed to us that the process of bacterial cloning led to great losses of DNA, probably in the step where bacteria were coaxed to take up DNA. Eddy, however, was convinced that the low efficiency seen in the cave-bear experiment was a fluke. He was characteristically enthusiastic about this in the phone conferences we had with his group. I was torn by our disagreement. It seemed, after many years of frustration, not only that it might be possible to arrive at a sequence of the entire Neanderthal genome but that there were multiple ways to do it. Yet I felt that the project would be tenable only if it required grams and not kilograms of bone, as Eddy's technique would require. 454's pyrosequencing seemed to fit the bill already, but eventually Eddy convinced me to give bacterial cloning another chance. So I decided to test the two approaches, bacterial cloning and direct sequencing of molecules, head to head, and to do it with the real thing: Neanderthal DNA.

We prepared two extracts from what we considered to be our best Neanderthal sample, a bone known as Vi-80. David Serre had sequenced a highly variable part of its mitochondrial DNA in 2004. In mid-October 2005, we sent off one extract to be directly sequenced by Michael Egholm and his crew at 454 Life Sciences and one extract to Eddy Rubin's group to be cloned in bacteria and then sequenced. The extracts had been prepared by Johannes Krause, working in our clean room. It was unnerving to then send them on to the labs in Connecticut and California where they might get contaminated. Once the tests proved which method was best, we would need to establish that method in our clean room.

Meanwhile, another new graduate student, Adrian Briggs, had arrived in our group. Fresh from undergraduate studies at Oxford, Adrian was the nephew of Richard Wrangham, the well-known Harvard primatologist. Both Adrian's family ties and his Oxbridge education had me worried that he would turn out to be snobbish and arrogant, but my fears were totally unfounded. Even better, Adrian had an amazing ability to think quantitatively about problems in a way no one else in our group did. Best of all, he never made the rest of us feel stupid, although he thought more quickly and accurately about problems than any of us. Whereas I had no more than a hunch that most of the DNA had been lost in the process of making the Berkeley cave-bear libraries, Adrian calculated that only about 0.5 percent of the cave-bear DNA we sent to Eddy Rubin's group had actually ended up in the bacterial libraries they produced. Adrian also calculated that in order to sequence the 3 billion–plus base pairs of a cave-bear or Neanderthal genome it would be necessary to isolate and sequence about 600 million bacterial clones, a logistical impossibility even at Eddy's Joint Genome Institute. It put my concerns about the cloning on a solid footing; obviously, the process of cloning in bacteria was nowhere near as efficient as would be needed to get the Neanderthal genome. In a rather tense phone conference in January 2006, Adrian presented these results to Eddy's group. Eddy, however, still felt that something had gone wrong with his laboratory's cave-bear libraries. In the meantime, work at both 454 and Eddy's lab went forward.

But we were not the only ones thinking of trying pyrosequencing on ancient DNA. Early in 2006, while Ed Green was busy analyzing our cave-bear and mammoth data, a paper appeared in *Science* by my previous graduate student, Hendrik Poinar, now at McMaster University in Ontario, in collaboration with Stephan Schuster at Penn State University. They used pyrosequencing applied directly to the DNA extract, just as we had done with 454 Life Sciences, to determine 28 million nucleotides of DNA from a permafrost mammoth.[6] I was happy that a previous student was doing this, even if our group was disappointed not to be the first to publish sequencing of ancient DNA using pyrosequencing. We had had our data from the mammoth and cave-bear bones for many months, but we'd spent a lot of time on two things the *Science* paper did not: analyzing how best to match the DNA sequences we had determined to reference genomes and considering how errors in the sequences would affect the results. Still, Hendrik's paper was more evidence that direct sequencing was the way to go. It also showed, once again, that permafrost material could be amazingly well preserved. About half of the DNA in Hendrik's sample was from the mammoth, a far cry from what we could hope for from our Neanderthals—we were happy when we had an extract containing 1 or 2 percent Neanderthal DNA. Hendrik's paper also illustrated a dilemma in science: doing all the analyses and experiments necessary to tell the complete story leaves you vulnerable to being beaten to press by those willing to publish a less complete story that nevertheless makes the major point you wanted to make. Even when you publish a better paper, you are seen as mopping up the details after someone who made the real breakthrough. Our group discussed this point extensively after Hendrik's paper appeared. Some felt we should have published earlier. In the end, the analysis of our cave-bear and mammoth sequences appeared in the September 2006 issue of *Proceedings*, where we, ironically, ended up reporting erroneous conclusions about deaminated A's giving rise to mutations in the sequences.[7]

In May each year, there is a meeting on genome biology at Cold Spring Harbor Laboratory on Long Island. This meeting is the unofficial summit of genome scientists from around the world, and presenters are expected to talk about novel and unpublished results. It tends to be an intense affair, colored by the rivalry among genome centers and sometimes by conflicts and aggressions carried over from the race to sequence the human genome.

In 2006, the genome meeting was even more intense for me than it normally is. We had just obtained Neanderthal sequencing results both from 454 Life Sciences and from Eddy Rubin's group at Berkeley, and we had done some preliminary analyses. I had two goals for my talk. First, I wanted to present the comparisons of the two different techniques to sequence ancient DNA. Second, I wanted to lay out a road map of how one might get to a whole-genome sequence of the Neanderthal and other extinct organisms. The results confirmed that direct pyrosequencing was the method of the future, so my emphasis would be on that.

I was unusually nervous when I arrived at Cold Spring Harbor. I was housed in a small spartan room on campus, an honor bestowed on frequent attendees of the meeting, while others have to be bused in from far-flung hotels. I spent the entire flight to New York, as well as the first night in my tiny room, preparing my talk. The next day I collected the people from my lab who attended the meeting and I gave a practice talk in a side hallway. I had the feeling that this would be a talk that would define what we were to do in the next few years.

It is rare to have the undivided attention of the audience when giving a scientific talk. The Genome Biology Meeting at Cold Spring Harbor is a case in point. I had given many talks there before and was used to watching most of the six hundred or so people in the room fiddle with their laptops as they checked through their own presentations or e-mailed colleagues— or dozed off due to the combined effects of jet lag and too many highly detailed talks. But this time was different. As I worked my way through the mammoth and cave-bear results toward the Neanderthal data, I could feel how I had the absolute and undivided attention of the audience. My last slide was a map of the human chromosomes, with little arrows showing where the tens of thousands of pieces of DNA we had sequenced from the Neanderthal matched. When the slide went up, I heard what sounded like a gasp from the audience. Our sequences added up to only about 0.0003 percent of the Neanderthal genome, but it was clear to everyone that we had shown that one could now—in principle—sequence the whole thing.

Chapter 11
Starting the Genome Project

That night, after returning to my little room at Cold Spring Harbor Laboratory, I lay down on the bed and stared at the ceiling. So far, I had had a nice—one might even say somewhat distinguished—career. I had a permanent research position with solid funding, was doing interesting projects, and got invited several times a year to give talks around the world. Now I had really stuck my neck out, publicly promising to sequence the Neanderthal genome. If we succeeded, it would clearly be my biggest achievement to date; but if we failed, it would be a very public embarrassment, almost surely a career-ending one. And I knew that succeeding would not be as easy as I had made it sound in my talk. We needed three things to succeed: many 454 sequencing machines, lots more money, and good Neanderthal bones. We had none of them, but fortunately no one else seemed to realize this. I knew it only too well, however, and I lay in bed a long time, with all the things we needed to make the project possible running through my head.

The first priority was access to lots of sequencing machines from 454 Life Sciences. An obvious move would be to visit Jonathan Rothberg in Branford, Connecticut, which was not far from Cold Spring Harbor. Next morning at breakfast, I collected the key people involved in our Neanderthal work, all of whom were at the meeting: Ed Green, Adrian Briggs, and Johannes Krause. After breakfast, we jumped into my rental car and took off for Branford. I have a deplorable tendency to pack too many commitments into too little time and as a consequence am chronically late for appointments, flights, and other scheduled activities. This outing was no exception. As we drove toward Port Jefferson on northern Long Island, we realized we were probably going to miss the ferry across the Long Island Sound to Bridgeport. As it happened, we were the last car to squeeze aboard (in fact, the rear of the car jutted out over the water as we steamed across). I hoped this close call was a good omen.

This was the first of what would be several visits to 454 Life Sciences. Jonathan Rothberg was just as intense and full of maverick ideas in person as he had been on the phone. For balance, there was Michael Egholm, the practical-minded Dane concerned with reality checks and getting things done. As the project progressed, I came to appreciate both men immensely; between Jonathan's vision and drive and Michael's down-to-earth practicality, they made a terrific pair. Our discussion that day was dominated by what it would take to sequence the Neanderthal genome. It was clear that we would apply the "shot-gun" technique that Craig Venter had introduced and used in his bid to sequence the human genome at his company Celera. This approach involved sequencing random fragments and then putting them together computationally by looking for overlaps between fragments. One major complication involves repetitive DNA sequences in the genome; such sequences make up about half of the genomes of humans and apes. Most of these repetitive sequences are a few hundred or even thousands of nucleotides long, and many occur not only a few but many thousands of times across the genome. Therefore, shot-gun approaches typically use not only short DNA fragments but also longer fragments so that one can "bridge" such repeat sequences with fragments that "anchor" it in single-copy sequences on each side of the repeat. This makes it possible to know where each repeat element sits in the genome. But our ancient DNA was already broken down into short pieces. Therefore, we planned to use the human reference genome (the first human genome, sequenced by the public genome project) as a template for reconstructing the Neanderthal sequences. But while this should work for DNA sequences that occurred a single time in the genome, we could not hope to determine the sequences of all the repetitive parts. To me, it seemed a small sacrifice: the single-copy sequences tend to be the most interesting parts of a genome as they contain the most genes with well-known functions.

We also needed to decide how much of the genome to sequence. Before visiting 454 Life Sciences, I had decided to sequence about 3 billion nucleotides from our Neanderthal bones. This goal was dictated mostly by what I thought was possible, and also because it was approximately the size of the human genome. The fragmented nature of the ancient DNA meant that we would get sequences of many bits of the genome just once; other bits twice, from two independent fragments; others three times; and so on. It also meant that there were many parts of the genome we would not see at all, simply because no DNA fragment we sequenced would happen to include them. Statistically, we could expect to get two-thirds of the entire genome at least once and so fail to see about one-third. In genome-speak, this is known as 1-fold coverage, since statistically each nucleotide has the chance to be seen once. I felt that 1-fold coverage was a feasible goal, and one that would provide a good overview of the Neanderthal genome. Importantly, the resulting genome would be a stepping stone of sorts. Future sequences, derived from other Neanderthals, could be put together with ours to arrive at higher "coverage" until eventually all of the genome, at least the parts that were not repetitive, had been seen.

The goal I had set ourselves was thus somewhat arbitrary. Compared with sequencing efforts expended on present-day genomes, it was also rather humble, as those other projects aimed for 20-fold coverage or more. However, the task was still monumental. Our very best extracts contained just 4 percent Neanderthal DNA. I counted on finding more such bones and hoped that some would contain even a bit more Neanderthal DNA, assuming that if the average stayed at 4 percent, to get our 3 billion nucleotides we would have to generate some 75 billion nucleotides in all. And since our fragments were short, 40 to 60 nucleotides on average, this added up to between 3,000 and 4,000 runs on the new sequencing machines. It was the equivalent of devoting the entire facility at 454 Life Sciences to the Neanderthal project for many months—and at normal customer prices, it would be impossible for us even to contemplate.

Ed, Adrian, Johannes, and I discussed all this with Jonathan and Michael. The project clearly had appeal not only to Jonathan but to 454 Life Sciences as a company, because of its potential both to provide truly unique insights into human evolution and also, more pragmatically, to give 454's technology even higher visibility. I gladly agreed that the company people would be real scientific partners as well as co-authors on future publications with us, but that didn't mean we could do sequencing for free. Finally, we arrived at a price: $5 million. I could not decide whether that was good or bad news. It was more money than I had hoped to pay, but not a completely outlandish sum. We said we would go back home and think about it.

After the negotiations were done, Jonathan offered the four of us takeout sandwiches and sodas and then asked if we wanted to see his house before we headed back to the meeting at Cold Spring Harbor. We agreed. After our late lunch, we followed him home. I had grown up in humble circumstances, and my mother, a refugee from the Soviet invasion of Estonia at the end of World War II, had transmitted a highly pragmatic outlook to me. As a result, I am not easily impressed by luxury. But the visit to

Jonathan's place turned out to be very memorable, even though we never got to see his house. Instead, we visited the grounds where he lived on a peninsula in Long Island Sound. On the beach, he had built an exact replica of Stonehenge—exact, that is, except that it was made of Norwegian granite and therefore heavier than the original, and it was slightly modified to account for how the sun would fall between the stones on the birthdays of his family members. As we walked among the huge monoliths, Jonathan turned to me and said, "Now you probably think I'm crazy." I of course said no, but not only out of politeness. I really didn't think Jonathan was crazy. He was deeply fascinated by ancient history, and more important, he had big ideas and was able to turn his dreams into reality. His Connecticut Stonehenge was, I thought, another good omen for our undertaking.

The next day, back at Cold Spring Harbor, I could not concentrate at all. Five million dollars was a lot of money, about ten times as much as a big research grant in Germany. The Max Planck Society provides generous funding to its institute directors so that they can concentrate on research and not on grant writing, but $5 million was still a much higher amount than the entire yearly budget for my department. I worried that we would need to turn this project over to some genome center, just because we didn't have the money. Then I remembered Herbert Jäckle, the developmental biologist who had helped persuade me to move to Germany in 1989 when he was professor of genetics in Munich. He, too, had moved to a Max Planck institute—the Institute for Biophysical Chemistry in Göttingen—and had again played an important yet unofficial role in getting me to move, in 1997, from Munich to Leipzig to join in establishing the Institute for Evolutionary Anthropology. In fact, ever since I had come to Germany, when I had faced crucial turning points in my scientific life, Herbert had always been there with support and advice. Now he was vice president of the biomedical section of the Max Planck Society. Fortunately, the society is a research organization where scientists, such as Herbert, rather than administrators or politicians are in charge. That very afternoon I decided to call him from Cold Spring Harbor.

I don't call Herbert often, so I think he realized that this was a matter of some import and urgency. When I got him on the line, I described our calculations on the feasibility of sequencing the Neanderthal genome and the cost, and I asked if he had any advice on how one might raise that much money in Europe. He said he would think about it and get back to me in a few days. I flew back to Leipzig the next day, torn between hope and despair. Perhaps we could find a rich benefactor, but how do you find one of those?

Two days after my return, true to his word, Herbert called me. The Max Planck Society, he said, had recently set up a Presidential Innovation Fund to support extraordinary research projects. He had discussed our project with the society's president, and the society was ready—in principle—to support us with the requested funds, distributed over three years. They had even reserved the money, pending a written proposal, which would need to be reviewed by experts in the field. I was totally taken aback; I don't recall even thanking him properly before hanging up. This money made all the difference in the world! I ran from my office into the lab and babbled the news to the first people I met. Then I immediately sat down and started drafting a proposal describing the results and calculations that had convinced us we could sequence the Neanderthal genome within three years, given sufficient resources.

At the proposal's conclusion, I had to present a financial plan. When I started working it up, something extremely embarrassing dawned on me. I had called Herbert from the United States and said we needed "5 million" for the project, thinking in terms of US dollars. Herbert, in Europe, must have thought I meant 5 million euros. He may even have *said* that the Max Planck Society had reserved "5 million euros" for our project, but I had been too excited to register it. At the exchange rate, then, that amounted to US$6 million. What to do? Perhaps I could quietly increase the budget to accommodate the additional 20 percent in funds—but that would be disingenuous and might even be noticed once we signed a contract with 454 Life Sciences. I called Herbert and, with considerable embarrassment, explained the situation. He laughed. Then he asked whether we might not have extra costs in Leipzig, above what we were to pay 454 Life Sciences for sequencing. Of course we would. We would have to extract DNA from many fossils to find good ones, and test them all by making sequencing runs from them ourselves. So we needed to buy our own sequencing machine from 454 for testing all the extracts, and we needed reagents to run it. With the difference the exchange rate made, we could really make the project fly. I was elated and wrote a plan that included the work that would be done at our institute in Leipzig.

Meanwhile, Eddy Rubin's group in Berkeley had made a bacterial library of the entire Neanderthal extract we had sent them. Jim Noonan, Eddy's postdoc, had sequenced every drop. What they had retrieved amounted to a bit over sixty-five thousand base pairs. In Branford, they had used about 7 percent of the extract we had sent them and produced about a million base pairs. So, just as Adrian had predicted, the direct-sequencing approach was about two hundred times more efficient in generating DNA sequences from an extract. Eddy insisted nonetheless that his method could be more efficient, and that we should continue to send extracts to him. This was a fundamental disagreement. I realized with some disquiet that I could no longer in good conscience send extracts to Berkeley, when we could generate so much more data from each extract in Branford. But I put the decision off, thinking that it would become obvious to Eddy that the bacterial cloning was inefficient once we wrote up a manuscript describing the results of the two different approaches.

However, by this point it was impossible to conceive of a way to write just one paper, given the use of two completely different methods, the tremendous difference in the amounts of data generated, and the disagreement with Eddy about the viability of the bacterial-library approach. So we decided to write two papers. One was to be written by Eddy with us as co-authors, the other by us and Michael Egholm, Jonathan Rothberg, and the others at 454. Eddy's paper stated: "The low coverage in library NE1 is more likely due to the quality of this particular library rather than being a general feature of ancient DNA," suggesting that if one assembled more libraries, better results would be achieved. Given that the earlier cave-bear libraries had been just as inefficient, I disagreed with the assessment, but we stayed civil. Eddy submitted the paper in June to *Science* and it was accepted in August. Because we had much more data to analyze for the 454 paper, we couldn't submit our paper until July to *Nature*. Eddy graciously arranged with *Science* to delay publication of the cloning paper until the paper with 454 Life Sciences had been reviewed and accepted in *Nature* so that the two papers could appear in the same week.

While this was going on, we began to prepare for what we hoped would be the production of large amounts of Neanderthal sequences. The first thing I did was to arrange production of 454 sequencing libraries in our clean room in Leipzig so that the precious, contamination-prone DNA extracts would not have to leave our laboratory. I also used a chunk of the new money to order a 454 sequencing machine so that we could test the libraries. Then Michael Egholm and I worked out a plan. We would make DNA extracts from bones, produce 454 sequencing libraries in our clean room, and use our new 454 sequencing machine to test the libraries. When we identified promising libraries, we would send them to Branford for production sequencing. The sequencing would be done in stages, and we would pay in installments once a certain amount of Neanderthal nucleotides had been sequenced. The latter was my suggestion, and I was amazed that 454 agreed to it, given that our earlier work together had shown that the best library so far had contained only 4 percent Neanderthal DNA and 96 percent assorted unwanted DNA of bacterial, fungal, and unknown origin. We did not yet know what percentage of Neanderthal DNA would be in the libraries we would produce. If it turned out to be 1 percent instead of 4 percent, then 454 would have to sequence four times as much to get its money, since the contract stipulated the number of *Neanderthal* nucleotides sequenced, not the *total* number of total nucleotides (which would include all the bacterial ones). Neither the scientists at 454 nor their attorneys who looked at the contract before it was signed appeared to take any notice of this. In a sense, it didn't matter, since there was a clause that allowed either party to get out of the collaboration at any time. We were obviously not going to be able to force 454 to sequence forever against their will. But it still seemed a much better contract than one that stipulated that the company would sequence a certain amount of raw nucleotides for us, irrespective of whether these were microbial or Neanderthal in origin.

I felt very good about the collaboration with 454. We complemented each other's strengths excellently, and the people at the company were fun and easy to talk to. However, one difference between us was that 454 was under great pressure to establish itself in an emerging market for high-throughput sequencing technologies that was clearly going to become very competitive. Already, two other big companies had announced their intention to start selling high-throughput sequencing machines. 454 therefore wanted positive publicity about their involvement in the Neanderthal project, and they wanted this publicity not in two or three years, when the Neanderthal genome would presumably be sequenced and published, but as soon as possible. Just as Michael Egholm took our concerns and priorities into account, I wanted to take their priorities seriously. So when the contract was signed with 454, we allowed them to arrange a press conference in our institute in Leipzig on July 20, 2006, shortly after we had submitted our joint paper to *Nature*. Michael and another senior executive from 454 flew in for the event. We also invited Ralf Schmitz, the curator of the Neanderthal type specimen who had given us samples from the Bonn museum in 1997. He brought along a copy of the Neanderthal bone from which we had determined the first Neanderthal mtDNA sequences. We wrote a press release that pointed out that we were putting together the methods for ancient DNA analysis that our group had developed over many years of painstaking work with 454 Life Sciences' novel high-throughput sequencing technology to analyze the Neanderthal genome. We also mentioned that, by coincidence, we announced this almost exactly on the day 150 years after the first Neanderthal fossil was discovered in Neander Valley.

The press conference was an electrifying event. The room was full of journalists, and media from across the globe followed it via the Internet. We declared that we would determine about 3 billion Neanderthal nucleotides within two years. It was wonderful to contemplate that what I had started secretly in the lab in Uppsala more than twenty years earlier, afraid that my PhD supervisor would find out what I was doing, had developed into this. It was a heady time.

It was also a time of great scientific and emotional ups and downs. About a month after the press conference came a definitive down. The two papers led by Eddy Rubin's and our group were not yet out, but we had already shared our 454 Neanderthal data with Jonathan Pritchard, a young and brilliant population geneticist at the University of Chicago who had helped Eddy analyze his smaller data set of cloned Neanderthal DNA fragments. We received an e-mail from two postdocs in Pritchard's group, Graham Coop and Sridhar Kudaravalli. They were worried about patterns they saw in the 454 data: in particular, there were higher numbers of differences from the human reference genome in the shorter DNA fragments than in the longer DNA fragments. Ed Green in our group quickly confirmed that they were right. This was worrying. It could mean that some of the longer fragments were not from the Neanderthal genome but represented modern human contamination. I e-mailed Eddy, telling him that we saw some worrying patterns in the 454 test data. We agreed to send our data to Eddy's group in exchange for their data. After the exchange of data, Jim Noonan in Eddy's group quickly e-mailed back and said that he saw what we and the Chicago postdocs had already seen in the 454 data.

It seemed that we might have to rewrite or withdraw our *Nature* paper, which was already in press, and I e-mailed Eddy, saying that we would try to figure out what was going on as fast as we could in order not to hold up his paper. Back when I was a postdoc in Allan Wilson's lab, we had once withdrawn a paper that *Nature* had already accepted because we had found that we had made a mistake in the analysis that changed the main conclusions we presented. I worried that we would have to do this again.

There was now frantic activity in our group. It was not unreasonable to assume that the patterns Jonathan's group saw were due to some level of contamination, but it was not straightforward to come up with an estimate of how much contamination there might be. It would have been an error to simply assume contamination was the problem, however. We were acutely aware that we did not understand many aspects of how the short, damaged ancient DNA sequences behaved in comparison with the human reference genome. Perhaps other factors than contamination were at play? Unfortunately, we needed to act fast as our paper was already in press and Eddy was eager to publish his paper.

Ed had noticed that the shorter Neanderthal fragments in our 454 data contained more G and C nucleotides than the long ones. G and C nucleotides tend to mutate more often than A and T nucleotides, so this could lead to more differences between present-day humans and Neanderthals in the short (and GC-rich) sequences than in the long (and AT-rich) sequences. To test this, Ed matched up short and long Neanderthal fragments to the corresponding sequences in the human reference genome and compared those sequences in the reference human genome to those from other present-day humans. Although those comparisons did not include any Neanderthal sequences at all, they nonetheless showed that the human sequences corresponding to the shorter Neanderthal sequences had more differences from other human sequences than the longer ones. This observation suggested that the GC-rich sequences simply mutated faster, so maybe it would account for the higher number of differences seen in the shorter sequences. Before we could be certain, however, other factors also needed to be considered, especially the way in which we mapped Neanderthal sequences to the human reference genome sequence. Ed noticed that longer fragments of Neanderthal DNA had a better chance of being matched in the correct position in the human genome than shorter fragments, simply because they contained more sequence information. Therefore, a higher percentage of the short fragments might actually be bacterial DNA fragments that just happened to be similar to some part of the human reference genome. This, then, also might contribute to the observation that the shorter fragments contained more differences from the human reference genome. Such a phenomenon might have been overlooked in other ancient data sets—for example, the mammoth data, where fragments were on average longer. But I felt very uneasy. It seemed that every day we uncovered new things about how short and long DNA fragments differed in terms of how they behaved in our analyses. Obviously, we did not understand everything that was going on. What's more, we still hadn't excluded the possibility that our samples were contaminated by modern human DNA.

We had, of course, considered the possibility of contamination from the outset. In the extracts we sent to Eddy and to 454, we had assayed the level of contamination based on mtDNA and found it to be low. We knew that contamination could have entered the extracts once they had left our laboratory; we had even put a caveat about this in our *Nature* manuscript. I felt strongly that the only solid assay for contamination we had was the one based on assessing the observed mtDNA fragments, since the mtDNA was the only part of the genome where we *knew* about differences between Neanderthals and modern humans. Everything else was influenced by imponderables, such as differences in GC content, differences in mismapped bacterial DNA fragments, and other unknown factors. So I argued that we should look again at the mitochondrial DNA in the sequences that had been determined by 454.

In 2004, we had sequenced a part of the mtDNA from the very same Neanderthal bone, Vi-80, from which we had prepared test extracts for 454 and Eddy's group. I suggested that we should look among the sequences we had gotten from 454. Surely some of those must overlap nucleotide positions that differed between this particular Neanderthal individual and present-day humans. This would tell us which fragments were unambiguously of Neanderthal origin and which were of modern human origin and would enable us to estimate directly the level of contamination in the actual final 454 data set. Frustratingly, Ed found that we did not have enough data in hand to do this. The sequences done by 454 contained only forty-one mtDNA fragments and none of them came from the part of the mtDNA genome that we had determined earlier from this or other Neanderthals. We checked the Berkeley data, but they were so scant that not even a single mtDNA fragment had been observed.

Happily, there was a solution: we had so much library left that we could simply sequence more DNA fragments. This should then yield fragments that could tell us whether we had contamination in the library or not. I contacted 454 and convinced the people there to quickly do more sequencing. They did six more runs in record speed, and as soon as the data were transferred to our server, Ed found six fragments that overlapped positions in the variable part of the mtDNA we had sequenced in 2004.

All six fragments matched the Neanderthal mtDNA and differed from present-day human mtDNA! These were direct data suggesting that we had very little contamination in our sequences. Interestingly, these molecules, although clearly ancient, were not particularly short; four of them were 80 or more nucleotides long. This suggested that truly ancient DNA fragments were present also among the longer DNA fragments. Thus it was likely that the differences seen between short and long molecules were due to factors other than contamination. Ed was so elated that he ended the e-mail to the group describing these results with "I could kiss every one of you."

We decided to go ahead with the *Nature* paper. Susan Ptak, a population geneticist in our group, sent a long technical e-mail to Eddy and Jim Noonan explaining why we felt that comparisons between long and short sequences were influenced by too many factors both known and unknown to represent strong evidence of contamination and explained why we trusted the direct mtDNA evidence more. She wrote: "Although there is indirect evidence which suggests some level of contamination, we now have a direct measure of the contamination rate in the final data set, which still suggests it is low." We received no reply to this e-mail. Given the rather tense relationship that had developed between our groups, we did not find this too surprising.

This was a tremendously stressful incident. Ironically, as it turned out, both Eddy and we were right. The future would show that the data generated at 454 did contain contamination, but also that the indirect ways of detecting contamination via comparisons of long and short fragments were largely inadequate.

The two papers were published in *Nature* and *Science* on the 16th and 17th of November.[1] There was the predictable excitement in the press, which I had by now gotten used to. In fact, I was much more preoccupied than excited. We had promised the world that we would sequence 3 billion base pairs of the Neanderthal genome within two years. Our paper ended with an estimate of what this would require—namely, about twenty grams of bone and six thousand runs on the 454 sequencing platform. We said that this was a daunting task, but added that technical improvements that would make the retrieval of DNA sequences on the order of ten times more efficient could "easily be envisioned." The improvements we had in mind involved losing less material when making libraries for sequencing and taking advantage of secret future improvements to the 454 machines that Michael had revealed to us.

Things were looking up, but a major challenge still remained: finding good Neanderthal bones. The truth was that we did *not* have anywhere near twenty grams of Neanderthal bone of the quality of Vi-80, the bone we used in the test runs for the two papers. In fact, the piece we had left from Vi-80 weighed less than half a gram. I optimistically told myself that since one of the first Vindija bones we tried contained almost 4 percent Neanderthal DNA, surely we would find others that were equally good. Perhaps we would even find some that were better. I had to turn my full attention to this problem as soon as possible. First, however, I had to undertake a more unpleasant task: ending the collaboration with Eddy Rubin.

Terminating a scientific collaboration is often difficult, and it is even more so when a collaborator has become a personal friend. I had stayed with Eddy's family in Berkeley; we had biked up the hills to his lab together; we had gone together to the theater in New York during Cold Spring Harbor meetings. I had always enjoyed his company. So I long pondered my e-mail to Eddy and wrote several drafts of it. I explained how I differed with him on bacterial cloning's usefulness, and said that I felt that our communication, particularly on this point, had not been productive. I also noted that it now seemed that his group was trying to do the same things our group was trying to do, rather than working in a complementary way. For example, in our phone conferences, they had suggested that we send them our DNA extracts and the PTB reagent we had synthesized so that they could treat our extracts with our PTB. Neither I nor my group had been thrilled by this notion. I hoped I had expressed my reasons for not working together in a way that wasn't hurtful or insulting, but it was still with some trepidation that I sent the e-mail. Eddy answered that he saw my points but that he continued to believe in the future potential for improvements and utility of bacterial libraries. I was relieved that he had taken my letter graciously, but we were now, clearly, competitors rather than collaborators.

The competition became apparent almost as soon as I turned my attention to the procurement of Neanderthal bones. Eddy was trying to obtain them too, I discovered, and from many of the same people we had worked with for years. In fact, I found out that already back in July, *Wired* magazine had published an article about Eddy's Neanderthal efforts. The *Wired* piece ended with a quote from Eddy: "I need to get more bone. I'll go to Russia with a pillowcase and an envelope full of euros and meet with guys who have big shoulder pads. Whatever it takes."

Chapter 12
Hard Bones

Even before our *Nature* paper came out, Johannes Krause had begun preparing extracts from Neanderthal bones we had collected from Croatia and elsewhere in Europe over the years, hoping to find a bone that might contain as much or more Neanderthal DNA as Vi-80. Johannes was tall and blond, not so dissimilar to the German stereotype. He was also very intelligent. He was born and had grown up in Leinefelde, the very same town where in 1803 Johann Carl Fuhlrott had been born. Fuhlrott was the naturalist who in 1857, two years before Darwin published *The Origin of Species,* had suggested that the bones found in Neanderthal derived from a prehistoric form of humans. This was the first time anyone suggested that other forms of humans had existed before current humans and Fuhlrott was widely ridiculed for his idea, but he would be proven correct when additional Neanderthals were unearthed. Fuhlrott became a professor at the University of Tübingen, where today, appropriately, Johannes is a professor.

Johannes had come to our department as an undergraduate specializing in biochemistry. It soon turned out that he was not only very good at bench work but had good judgment and comprehension of all the complex experiments going on in the group. I always enjoyed talking with him, but as the months passed he seemed to bring me only bad news. None of the many extracts he prepared from various Neanderthal bones contained anything like the amount of Neanderthal DNA we had seen in Vi-80. Most of them contained no Neanderthal DNA at all, or so little that he could barely detect Neanderthal mtDNA by means of the PCR. We urgently needed more and better bones.

The obvious place to go was back to the Institute for Quaternary Paleontology and Geology in Zagreb, where the Vindija collections, including the remainder of the Vi-80 bone, were housed. In April 2006, I had written to the Zagreb institute. I said that we were interested in sampling the bone we called Vi-80[1] again and perhaps other bones excavated between

1974 and 1986 by Mirko Malez in Vindija Cave. Sadly, I learned that Maja Paunovic, with whom I had worked in 1999, had died. There was now no paleontologist in charge of the collection. The head of the institute was Milan Herak, an emeritus professor of geology at the University of Zagreb, who was eighty-nine and rarely if ever visited it. An elderly lady by the name of Dejana Brajković did the day-to-day work, together with Jadranka Lenardic, her younger assistant. I wrote a letter to both women, explaining that we would like to continue our successful collaboration on the Vindija collection—a collaboration that had already resulted in three high-profile publications. I suggested visiting them to discuss this and perhaps sample a few more of the bones. We agreed that I would visit Zagreb and give a seminar at the university on our work. But in May 2006, four days before Johannes and I were to leave for Zagreb, I received an e-mail saying that it would be impossible to sample any of the Vindija bones. The bones had to be "registered," they said, and only after that event, at some undetermined time in the future, would it become possible to work with the bones. I sensed that someone else was behind this sudden turn of events. Their letter mentioned Jakov Radovčić, a famous paleontologist who curates the huge collection of much older Neanderthal bones found at Krapina housed in the Croatian Museum of Natural History in Zagreb. Although he had no formal authority over the Vindija collection, which belongs to the Croatian Academy of Sciences and Arts, I suspected that he wielded enough unofficial influence over the two women at the institute to have interfered with our arrangements. Still, I decided not to take no for an answer and go anyway. It seemed to me that the scientific promise of our project would be enough to persuade the people in Zagreb that our work should proceed.

Johannes and I arrived in Zagreb in early June and went directly to the institute, where I had spent considerable time with the late Maja Paunović several years before. It was still a rather dusty place, not exactly bristling with energy. Dejana Brajković and her assistant seemed nervous about our visit. They refused to let us see, let alone sample, the specimens and said that we would have to consult the Academy of Sciences and Arts before doing so. But after drinking coffee and chatting with them for a while, we were at least allowed to look at the bones. Some parts of the collection were in disarray, which may have contributed to their reluctance to allow us to work with it. I felt that establishing a proper catalog of the bones was a very good idea indeed. I was particularly attracted to a box of bones that Tim White, a well-known paleontologist at UC Berkeley, had set aside when he studied the collection a few years earlier. It contained fragments of bones that the excavator Mirko Malez had thought were from cave bears but that Tim thought could potentially stem from Neanderthals.

Looking at these bone fragments, I was reminded of something Tim had mentioned to me when we met at Berkeley a year earlier. The Vindija Neanderthal bones—all of them—were crushed into small fragments. This is typical of many, even most, sites where Neanderthal bones are found. Of course, it is not surprising that bones thousands of years old are not in good condition. But there are often cut marks on the bones where muscles and tendons had been attached as well as cut marks on the skulls. In short, the skeletons had clearly been deliberately de-fleshed, and bones containing marrow had been crushed, presumably to get to their nutritious contents. Tim had pointed out to me the similarity of this pattern of Neanderthal bone fragmentation to a gruesome Anasazi site from the American Southwest, where around AD 1100 some thirty men, women, and children had been butchered and cooked. He told me that the way in which many Neanderthal bones were crushed was similar to the way the bones of animals, such as deer, that were butchered by Neanderthals were crushed (see Figure 12.1). We will probably never know how common it was for Neanderthals to kill and eat other Neanderthals, or, indeed, whether these Neanderthal corpses might have been butchered and perhaps eaten as part of some mortuary ritual. But given that Neanderthal skeletons are found

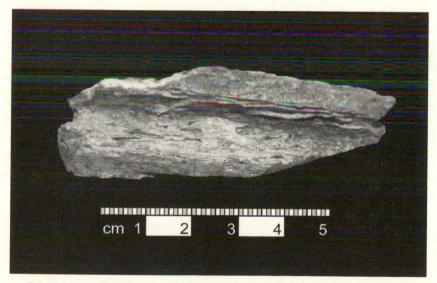

FIGURE 12.1. The bone 33.16 from Vindija Cave that we used for sequencing the Neanderthal genome. It has been crushed, presumably to get to the nutritious marrow. Photo: Christine Verna, MPI-EVA.

intact at some sites, and sometimes even positioned in ways that suggest deliberate burial, it seems likely that the Neanderthals in Vindija Cave had been unlucky enough to run into hungry neighbors.

Oddly, the fact that the Vindija Neanderthals were cannibalized, or at least de-fleshed, by other Neanderthals may be responsible for the fact that we had found rather a lot of Neanderthal DNA and relatively little bacterial DNA in at least some of the Vindija bone fragments. If the Neanderthal corpses had been buried, months would have passed before all its soft tissues were consumed by bacteria and other microorganisms. There would thus be ample time for bacteria to penetrate the bones, degrade the Neanderthal cells and their DNA, multiply, and eventually die themselves. Extracting DNA from such a bone would mostly yield DNA of microorganisms. If, on the other hand, the Neanderthal had been butchered, the bones crushed, gnawed, and sucked free of any meat and marrow before being tossed aside, some bone fragments would quickly dry out, limiting the chance for bacteria to multiply in them. Thus, we might have Neanderthal cannibalism to thank for the success of retrieving DNA in some specimens from Vindija.

All this went through my head as I looked in the box containing bones so crushed that it was impossible to tell whether they came from animals or Neanderthals. I turned to Dejana Brajković and asked if we could at least sample some of these fragments, whose source was so ambiguous. I argued that if DNA were preserved in them, we could determine what species they came from. But Brajković was adamant; we could not touch any of the bones. She said she had heard that in a few years one would be able to hold a sensor close to a bone and thereby determine its entire genome sequence; thus it was inadvisable to sacrifice even a tiny part of a bone fragment now. I agreed that techniques would certainly improve in the future but expressed polite doubt about whether we would live to see the advance she envisioned. Again, I suspected the influence of powers greater than hers. I said we would discuss our needs with the Croatian Academy and be in touch.

In the afternoon, we visited Jakov Radovčić at the Museum of Natural History. He appeared to be supportive of our project but expressed grave reservations about the sampling of any bones in either the Krapina or Vindija collections. I was sure we had not yet gotten to the bottom of things, and in a gloomy mood we returned to our small and scruffy hotel room. I lay on the bed, gazing at the paint peeling from the ceiling and feeling completely frustrated. As far as I knew, these were the bones containing the best Neanderthal DNA in the world. Many of them were of little or no morphological value, so fragmented that you couldn't even tell whether they came from a Neanderthal, a cave bear, or some other animal. Yet some unknown person with influence over the people at the institute was apparently determined to make it impossible for us to work on them. Like a child denied his favorite candy, I felt like screaming and kicking, but my Swedish upbringing kept me from venting in such an obvious way. Instead, Johannes and I spent the evening in a bad restaurant around the corner from our hotel, brooding about our mysterious enemy.

The next day I gave a talk before the medical faculty of the University of Zagreb about ancient DNA in general and our Neanderthal work in particular. It was well attended, and many of the students asked questions. It cheered me a bit that some young people in Zagreb were enthusiastic about science. In the evening, we had dinner with Pavao Rudan, a professor of anthropology at the university, who stems from an old family of landowners on the beautiful island of Hvar, off the Adriatic coast. He invited us to join him and his colleagues at a restaurant named Gallo, which turned out to be one of the best restaurants I have ever been to. Course after course of excellent seafood and creative Mediterranean dishes were served, along with good wines. The meal was rounded off by a wonderfully refreshing drink of fruit juice, champagne, and some other ingredient that I couldn't make out. I felt slightly better. Then, Pavao started to talk about science. As it happened, my conversation with him was to improve my spirits in a much more lasting way than the exquisite dinner.

First, we talked about his work on small human populations on the Croatian islands. He was trying to find genes and lifestyle traits that contribute to common disorders, such as high blood pressure and heart disease. For many years, he had had grants from the United States, France and the UK for this project, which testified to his scientific credentials. I figured that he would recognize a good project when he heard about it, so I talked at length about our plans and our problems. Pavao listened to my plight with sympathy and was willing to help. Crucially, he knew how to navigate the Byzantine politics of Croatia. He told me that he had just been elected to the Croatian Academy of Sciences and Arts and would soon be inducted as a member. He suggested that we approach the project not just as a collaboration between our research group and the Zagreb institute where the Vindija collection was housed but as a collaboration between the Croatian Academy and another academy—that is, one I belonged to as a member.

I was in fact a member of several scholarly academies. Such memberships are honors that I had until then regarded as quite irrelevant to my everyday science. I never attended their meetings, having imagined that they consisted of earnest discussions attended by esteemed scientists well into their dotage. But now they suddenly seemed important. Which academy should we approach? I suggested the National Academy of Sciences in the United States, perhaps the most prestigious academy membership I held, but Pavao advised against it. He suggested we should rather approach some academy in Germany. We settled on the Berlin-Brandenburg Academy of Sciences and Humanities, of which I had been a member since 1999. He suggested that I approach the president of the Berlin Academy and ask him to write to the president of the Croatian Academy to propose our project as a collaboration. He also advised me to wait a few weeks, until he had been inducted into the Croatian Academy. Along with other sympathetic members, he could then put in a good word for the project with its president.

The next morning, Johannes and I flew back to Leipzig. I was feeling a bit more optimistic. We didn't have the bones with us, as I had hoped, and persuading the Croatian Academy that it would be in the best interests of science to work with us remained a challenge. But with the help of Pavao, we might have a chance.

Back home, I immediately phoned Günter Stock, president of the Berlin-Brandenburg Academy. He listened intently and was ready to help; he liked the idea of strengthening ties with Croatia. With the help of his assistant for foreign relations, I drafted a letter from him to the president of the Croatian Academy, proposing the Neanderthal Genome Project as a collaboration between the two academies. We also suggested that we would be willing to support the establishment of a catalog of the Vindija collection, by donating a computer and resources for someone to do the work.

But I didn't leave it at that; I wanted to do everything I could to overcome the mysterious resistance in Zagreb. One way to do this would be to involve all relevant parties there in the project. So I wrote to Jakov Radovčić and invited him to the upcoming July press conference with 454, suggesting that he present the paleontological aspects of Neanderthals to the press. He responded that he had other obligations that made it impossible for him to attend. I also contacted Frank Gannon, director of EMBO, the European Molecular Biology Organization, to which I also belonged, and asked him to contact Dragan Primorac, the Croatian minister of Science, Education and Sports, on our behalf. Dragan Primorac is an unlikely politician. He is a professor of forensic science at the University of Split in Croatia as well as an adjunct associate professor at Penn State University in the United States. Dragan, who has since become a friend, answered that he would put in a good word for our project with the academy. I had no idea whether all these initiatives would help our project, but I wanted to leave no stone unturned.

In the meantime, the letter from Professor Stock on behalf of the Berlin-Brandenburg Academy formally proposing the Neanderthal project, as well as a letter from me, had arrived at the academy in Zagreb. Pavao Rudan, asked by his colleagues for his opinion, suggested some conditions for the proposed collaboration: at least one Croatian co-author should appear on all papers we were to publish on the Vindija material, the Croatian Academy should be mentioned in the acknowledgments, and at least two scientists from Croatia should be invited to Leipzig each year as long as the project lasted. I agreed to these conditions and added that we, along with the Berlin Academy, would support the establishment of a catalog of the Vindija collection.

All this took time. Summer turned into fall, and fall turned into winter. I was meanwhile pursuing other promising Neanderthal sites, concentrating on places where our previous work had shown that DNA was preserved. The first and most obvious one was the Neander Valley site itself, where the type specimen had been found in 1856. At that time, the cave had not been scientifically excavated but was emptied out by quarry workers, who collected bones as they happened to notice them. Since then, the entire cave, as well as the small mountain where it was situated, had been quarried away for limestone. Frustratingly, many of the bones of the type specimen had never been collected. Some years earlier, Ralf Schmitz, with whom we had worked on the type specimen, had had the crazy but brilliant idea of trying to find the missing bones. After painstaking investigation of old maps, long walks in Neander Valley, and the exercise of a great deal of intuition, he had managed to find the place, then partly under a garage and car-repair shop, where much of the cave debris had been deposited 150 years ago. He began an excavation, and his efforts paid off handsomely in that he found not only fragments of the type-specimen individual but also bones from a second individual. In 2002, we retrieved mtDNA from this individual and published it with Ralf.[2] Now Johannes returned to the bits of samples we had left and did new DNA extractions, analyzing them with our new methods to look for nuclear DNA. The results were discouraging. The extracts contained between 0.2 percent and 0.5 percent Neanderthal DNA—not enough to sequence the genome.

Another site, Mezmaiskaya Cave in the northwestern Caucasus, had been excavated by an archaeologist couple, Lubov Golovanova and Vladimir Doronichev, who were based in St. Petersburg, Russia. They had found remains of a small Neanderthal child. This child had probably been deliberately buried, and not consumed, as all its bones were intact and found in their expected positions. An exciting aspect of this child was that while the Neanderthals we had analyzed up to that point were all about 40,000 years old, this baby was between 60,000 and 70,000 years old. Lubov and Vladimir had visited our institute, bringing with them a small piece of a rib from the child for us to analyze, and also a fragment of a Neanderthal skull found in a higher layer of the cave. When Johannes made extracts from these specimens, the rib turned out to contain 1.5 percent Neanderthal DNA. This was still not as much as we had hoped for; moreover, the rib was so tiny that we could never hope to get enough DNA for a genome sequence from it. But it might contribute some data to our efforts.

The third site we explored—El Sidrón—was in Asturias, in northwestern Spain. I visited it in September 2007. When a child dreams of becoming a paleontologist, this is the type of site he or she imagines. El Sidrón is located in beautiful countryside. The cave entrance is small and hidden, and the cave has served as a refuge for people throughout the ages. In front of the entrance is a memorial to a fighter who hid there during the Spanish Civil War and was killed by the fascists. After crawling through the entrance, one walks about two hundred meters to a side gallery, twenty-eight meters long and twelve meters wide, on the right. There, Professor Marco de la Rasilla of the University of Oviedo and his collaborators and students excavate every summer. They have found bones from one Neanderthal infant, one juvenile, two adolescents, and four young adults. The long bones were crushed and full of cut marks. Only the bones of the hands were found together; they had been separated from the bodies and thrown to the side. Marco de la Rasilla believes that the body parts were disposed of in a small pond on the surface some 43,000 years ago and then washed into the cave.

New bones were being discovered at this site each summer, and we agreed that they should be collected for DNA analysis in ways that would maximize DNA preservation and minimize the chance of contamination by present-day human DNA. Working with Carles Lalueza-Fox, a molecular biologist at the University of Barcelona, and Antonio Rosas, a physical anthropologist at the National Museum of Natural Sciences in Madrid, the excavators equipped themselves with sterile gloves, clothing, face shields, and other tools typically used in our clean room. When they came across bones deemed suitable for DNA extraction, they donned the sterile outfits, removed the bones, and placed them directly in an icebox to freeze them. Back in Antonio's lab in Madrid, computer tomography of the bones was performed to document their morphology. The bones were then sent, still frozen, to us in Leipzig. Almost no one had touched them since their discovery, and bacterial growth would also have been minimized. I had high hopes that when Johannes made extracts, they would contain a lot of Neanderthal DNA, but of all the DNA in the bones, just 0.1 to 0.4 percent came from Neanderthals. From none of these sites, nor several others that we tried with even less luck, did we find enough DNA for a Neanderthal genome sequence. Vindija Cave was the only site where we had so far found a bone that had anywhere near enough DNA. Yet in Zagreb, things moved at a glacial pace, if at all.

One bright spot, late in the summer of 2006, was the arrival of a talented Croatian graduate student, Tomislav Maričić, in our group. Tomi had accompanied us when we visited the Institute for Quaternary Paleontology and Geology, and his cultural connections with Croatia came in handy as we tried to reach an agreement over the Croatian Neanderthals. Our project had become a matter of public debate there—a debate I could follow thanks to Tomi's translations of the Croatian newspapers. In July, after we had announced the Neanderthal Genome Project at the press conference in Leipzig, one of the big dailies, *Jutarnji List,* interviewed Jakov Radovčić, who was described as someone "without whom no Neanderthal research can be imagined." Said Jakov: "The question is: What is the goal of the research? Also, it is still not clear that one can retrieve the whole Neanderthal genome. . . . They are using a chemically aggressive method which destroys the material, which is too precious for us to sacrifice." In November the same paper quoted him again: "Three and a half months ago, Svante Pääbo

was in Zagreb looking for more samples for his molecular genetic analysis. . . . However, I think we should take special care of the samples and
keep them safe, so the next generation of researchers can use them."

Prompted by these remarks, I sent Jakov a long, polite e-mail, in
which I once again explained our project. He answered that after some cu-
ratorial formalities that were likely to take "several weeks or a few months,"
he would "strongly support" our project. Meanwhile, rumors were flying hither and yon in Zagreb. It was frustratingly unclear to me just who
supported and who was against the project, what was said by whom, and
whether people really meant what they said to me directly. The only people in whom I had solid confidence were Pavao Rudan and two friends of
his, both members of the Croatian Academy of Sciences and Arts, who
supported us. One of these was Željko Kučan, a statesman-like scientist of
poise and judgment who had been the first to introduce the study of DNA
at Zagreb University, some fifty years earlier. The other was a geologist
named Ivan Gušić, known as "Johnny" to his friends. Jovial, positive, and
always friendly, Johnny was soon to become the new head of the Institute
for Quaternary Paleontology and Geology (see Figure 12.2).

In late November, Pavao used the occasion of the publication of our
Nature and *Science* papers to take a public stand in our favor. He wrote
an article about the project in the Sunday edition of *Vjesnik,* the Croatian
newspaper of record, emphasizing that DNA studies could reveal much
about human evolution and that the Vindija material was essential for this.
"Therefore, the cooperation with the Max Planck colleagues should be continued and made stronger," he argued. "It is the Vindija samples that are
kept in a HAZU [the Croatian Academy's acronym] collection that can
make it possible to retrieve a Pleistocene hominid genome for the first
time in history. . . . Future cooperation between HAZU and [the] Berlin-
Brandenburg Academy, especially with Svante Pääbo's team, will improve
paleoanthropological, molecular genetics, and anthropological science." I
very much hoped that our work would eventually show that Pavao had not
misplaced his trust.

Slowly the Croatian tide turned in our favor. On December 8, 2006,
after many vicissitudes, most of them incomprehensible, a memorandum
of understanding between the Zagreb and Berlin academies was signed.
What a relief! Finally nothing stood between us and the bones. As soon as
was feasible, I arranged to visit Zagreb with Johannes and Christine Verna,
a young French paleontologist from the Department of Human Evolution in our institute in Leipzig. She was to spend ten days at the Institute

FIGURE 12.2. Pavao Rudan, Željko Kučan, and Ivan "Johnny" Gušić, the three members of the Croatian Academy of Sciences who made it possible for us to sample the Neanderthal bones from Vindija Cave. Photo: P. Rudan, HAZU.

for Quaternary Paleontology and Geology making a preliminary catalog of all the Neanderthal bones in the Vindija collection. Johannes and I spent four days in Zagreb and then returned to Leipzig in the company of Pavao, Željko, and Johnny, who carried eight bones from Vindija in sterile bags, including the celebrated Vi-80, now officially known as Vi-33.16 (see Fig. 12.1).

We arrived late at night. The first thing we did next morning was to bring the bones to the Department of Human Evolution, where, still in their bags, they were scanned by computer tomography so that their morphology would be forever preserved in a digital form. Then the bones went into the clean room, and Johannes took over.

Using a dental drill with a sterilized bit, he removed two or three square millimeters of the surface from each bone. Then he drilled a small hole into the compact part of each bone, pausing frequently to avoid heating the bone and potentially damaging the DNA (see Figure 12.3). He collected about 0.2 grams of bone, adding it to a solution that within a few

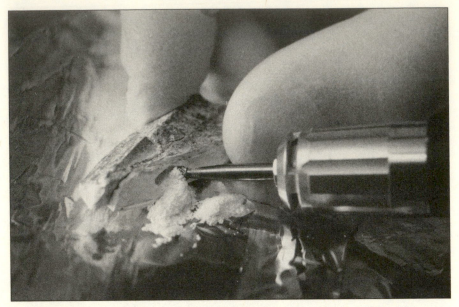

FIGURE 12.3. Sampling a Neanderthal bone with a sterile drill. Photo: MPI-EVA.

hours bound the bone's calcium. What was then left of the bone was a pellet of proteins and other components from its nonmineral portion. The DNA, however, was in the dissolved liquid part, and Johannes purified it by letting it bind to silica—the technique that Matthias Höss, fourteen years earlier, had found to be particularly good at isolating DNA from ancient bones.

To make the DNA molecules amenable to 454 sequencing, Johannes used enzymes to fill in and chew away any unraveled single-stranded DNA at the ends of molecules. That enabled him to use a second enzyme to fuse short synthetic pieces of modern DNA, called adaptors, to the ends of the ancient DNA. After adaptors have been added to DNA molecules, they can be "read" by sequencing machines just like books, so the collection of them is called a library. The adaptors had been synthesized especially for this project, and they contained a short additional sequence of four bases, TGAC, positioned so that it would abut the ancient fragments as a kind of marker or tag. This was one of those small technical details that often make a huge difference in molecular biology in general and ancient DNA research in particular. We had introduced these tags because our ancient DNA library had to leave the clean room to be sequenced on the 454 machine. In order to ensure that DNA from other libraries in our lab could not somehow end up in the Neanderthal libraries, we used these special adaptors and trusted only sequences that started with TGAC. We described this adaptor innovation in a 2007 paper.[3]

Using these procedures, Johannes prepared extracts and libraries from the eight new Vindija bones. He then used the PCR to see if there was Neanderthal mtDNA in the extracts and to estimate the extent of modern human contamination. Nearly all of the bones contained Neanderthal mtDNA. This was encouraging, but after our disappointments with the bones from Russia, Germany, and Spain, I did not allow myself to get enthusiastic. We immediately sequenced a sample of random DNA fragments from each of the libraries to estimate their proportion of nuclear Neanderthal DNA, and for the few days it took to get the results, I could hardly concentrate on any of my other work. We had announced to the world that we would sequence the Neanderthal genome. If these new Vindija bones did not contain enough nuclear DNA to do so, I was certain that we would have to announce our failure. I did not know where to look for any better bones.

When the results were in, they showed that some of the bones contained between 0.06 and 0.2 percent Neanderthal nuclear DNA, similar to what we had seen at the other sites. But three bones contained more than 1 percent Neanderthal nuclear DNA, and one contained nearly 3 percent. This was our favorite bone Vi-33.16, aka Vi-80. We had not found the magical bone with huge amounts of nuclear Neanderthal DNA that we had hoped for, but it was a bone we could work with.

All was not lost.

Chapter 13

The Devil in the Details

Over the Christmas and New Year holidays I used my free time to pon-
der our situation. It was sobering. When I calculated how much of such
bones we would need to complete the genome sequencing I came to tens
of grams, more than the entire weight of the bones we had. I felt very bad.
Had I been hopelessly overoptimistic, or naïve, to think that we could do
it? Had I been foolhardy to think that we would find a bone from Vindija
Cave that contained more Neanderthal DNA than the first ones we had analyzed? Had I put too much belief in 454 magically coming up with more
powerful sequencing machines that would somehow allow us to sequence
more? Why had I risked my calm and orderly scientific life on this gamble,
which it now seemed I was more than likely to lose?

My twenty-five years in molecular biology had essentially been a continuous technical revolution. I had seen DNA sequencing machines come
on the market that rendered into an overnight task the toils that took me
days and weeks as a graduate student. I had seen cumbersome cloning of
DNA in bacteria be replaced by the PCR, which in hours achieved what
had earlier taken weeks or months to do. Perhaps that was what had led
me to think that within a year or two we would be able to sequence three
thousand times more DNA than what we had presented in the proof-of-
principle paper in *Nature*. Then again, why wouldn't the technological revolution continue? I had learned over the years that unless a person was
very, very smart, breakthroughs were best sought when coupled to big improvements in technologies. But that didn't mean we were simply prisoners
awaiting rescue by the next technical revolution. Perhaps, I thought, we
could help the technology along a bit.

I reasoned that since we had so little bone, and since it contained so
little DNA, we needed to minimize losses of DNA from extraction to library. At our first Friday meeting after the holiday breaks I tried to instill
a sense of acute crisis in the group. I said that it was now clear that we

were not going to find a magical bone that would save us by containing lots of Neanderthal DNA. We needed to make do with what we had, and that meant rethinking every single step we did in the lab. I argued that the losses were probably huge. For example, the procedures used to purify DNA produced solutions that contained only minute quantities of other components, such as proteins. But the price of such purity is the loss of much DNA. If we could minimize such losses, perhaps the bones we had would be enough—at least once 454 Life Sciences finally had its new, more efficient machines ready.

I cross-examined my group week after week, asking repeatedly about every step they did in the lab. Perhaps the strategy of returning to the same questions over and over again was something I had retained from my training as an interrogator of prisoners of war during my otherwise long-forgotten military training in Sweden in my youth. The more I asked, the more I came to suspect that the recommended 454 protocols for pre-paring sequencing libraries, which were heavy on purification, might lead to undue loss of DNA. I insisted that we systematically analyze each step. How could we best do that?

When I was a graduate student the use of radioactivity was central to almost every experiment in molecular biology, but the cumbersome safety precautions required had long since inspired biologists to use nonradio-active assays. As a consequence, biology students today have almost no experience in working with radioactivity. However, radioactive labeling remains one of the most sensitive ways to detect tiny amounts of DNA. So in one of our Friday meetings, I suggested that Tomi Maricic label a small amount of DNA with radioactive phosphorus and use it to prepare a sequencing library. He could then collect the side fractions that were nor-mally discarded and measure how radioactive they were. The amount of radioactivity he detected in a side fraction would directly measure the loss of DNA at that step.

I assumed that the silence which greeted this idea in our Friday meeting was a tribute to the quiet elegance of this approach. But really I had run headlong up against an aspect of how our group functions. This aspect is, I believe, one of its biggest strengths, but at times it proves a weakness. I have encouraged a culture in which all ideas are discussed; everyone in the meeting is expected to speak his or her mind, and in the end we try to reach a consensus about what should be done. But as in any democracy, irrational ideas can sometimes win the day. My radioactiv-ity plan aroused skepticism in several people who were influential in the group. They made a number of objections to it, fueled (I thought) by an unconscious reluctance to adopt a method they had little experience with and which sounded old-fashioned and unsafe, if not downright scary. I decided not to force the issue. Other methods were tried instead, such as measuring the DNA amounts in each step of the library preparation and using more modern PCR-based approaches. But these methods either were not sensitive enough or proved ineffective in other ways. Over the next few months, I continued to suggest the radioactivity experiment, with increasing impatience, at times longing for a return to a more autocratic era, when the professor's word was law. Yet still I acquiesced, not wanting to put a chill on the free-floating exchange of ideas I felt was so valuable in the group.

Finally, when all other efforts had come to naught, it was the group that acquiesced. Tomi reluctantly ordered some radioactive phosphorus, labeled some ordinary human DNA we used for test purposes, and took it through the steps of preparing a 454 sequencing library. The results were stunning. He showed that in each of the first three major steps in the preparation, between 15 and 60 percent of the DNA was lost—a level not entirely unexpected in a biochemical separation. But in the last step, where the complementary DNA strands were separated with a strong alkaline solution, more than 95 percent of the input DNA was lost! Others who used this separation method with ordinary modern DNA had not noticed its inefficiency, because they had so much DNA that these enormous losses didn't matter to them. For our ancient work, though, they were catastrophic. Once the problem was identified, a simple remedy was devised. Alkaline solutions are not the only way to separate DNA strands; they also separate when they are heated. So Tomi tried heating and found from 10 to 250 times more radioactivity in the final DNA preparation! This was a great, indeed game-changing, advance.

Most labs discard side fractions as by-products. Fortunately, we had saved all of ours from our previous experiments. For years I had insisted on doing so, just in case something came along that would make them useful. This was easily one of my least popular ideas and caused many freezers to be filled with frozen side fractions that no one thought would ever be used. But thankfully in this case the crazy idea of the professor had been adhered to by the group. So now Tomi could simply heat the side fractions from earlier library preparations from the Vindija bones and retrieve additional, relatively copious amounts of Neanderthal DNA without even having to do any more extractions. He also optimized other steps in the library preparation. These changes resulted in a protocol several hundred times more efficient in turning the extracted DNA into a library ready for sequencing.[1]

Following consultation with our Croatian partners, we dedicated three Vindija bones—Vi-33.16 along with two new bones, Vi-33.25 and Vi-33.26—to the project. All seemed to be fragments of long bones that had apparently been crushed to get at the marrow (see Figure 12.1). Thanks to Tomi's advance we could now in principle produce libraries that contained 3 billion nucleotides of Neanderthal DNA from just these three bones. But the libraries would still contain at least 97 percent bacterial DNA, so the people in Branford would need to do between four thousand and six thousand runs on their sequencing machines to arrive at 3 billion base pairs of Neanderthal DNA. This was far more than we could ever imagine convincing Michael Egholm to do.

It seemed to me we were still stuck, until someone suggested that perhaps we could find pockets in our three bones where they contained much less bacterial DNA and therefore, relatively speaking, more Neanderthal DNA. Now and again over the years we had indeed seen indications that some parts of a bone might contain higher amounts of bacterial DNA than others, perhaps because bacteria had found growth conditions better in one part of the bone, and therefore multiplied more there, than in other parts. So, fueled by this hope, Johannes tried to systematically identify the best regions to sample. He drilled the bones until they looked first like flutes and then like Swiss cheese. He did indeed find a 10-fold difference in the percentage of Neanderthal DNA in regions just a centimeter or two apart, but the best regions still contained no more than 4 percent Neanderthal DNA!

We came back to this problem again and again in our Friday meetings. To me, these meetings were absorbing social and intellectual experiences: graduate students and postdocs know that their careers depend on the results they achieve and the papers they publish, so there is always a certain amount of jockeying for opportunity to do the key experiments and to avoid doing those that may serve the group's aim but will probably not result in prominent authorship on an important publication. I had become used to the idea that budding scientists were largely driven by self-interest, and I recognized that my function was to strike a balance between what was good for someone's career and what was necessary for

FIGURE 13.1. The Neanderthal genome group in Leipzig 2010. From the left: Adrian Briggs, Hernan Burbano, Matthias Meyer, Anja Heinze, Jesse Dabney, Kay Prüfer, me, a reconstructed Neanderthal skeleton, Janet Kelso, Tomi Maricic, Qiaomei Fu, Udo Stenzel, Johannes Krause, Martin Kircher. Photo: MPI-EVA.

a project, weighing individual abilities in this regard. As the Neanderthal crisis loomed over the group, however, I was amazed to see how readily the self-centered dynamic gave way to a more group-centered one. The group was functioning as a unit, with everyone eagerly volunteering for thankless and laborious chores that would advance the project regardless of whether such chores would bring any personal glory. There was a strong sense of common purpose in what all felt was a historic endeavor. I felt we had the perfect team (see Figure 13.1). In my more sentimental moments, I felt a love for each and every person around the table. This made the feeling that we'd achieved no progress all the more bitter.

During the spring of 2007, the Friday meetings continued to show our cohesive group from its best side. People threw out one crazy idea after another for increasing the proportion of Neanderthal DNA or finding microscopic pockets in the bones where preservation might be better. It was almost impossible to say who came up with which idea, because the ideas were generated in real-time, during continuous discussions to which everyone contributed. We started talking about ways to separate the bacterial DNA in our extracts from the endogenous Neanderthal DNA: maybe the bacterial DNA differed from the Neanderthal DNA in some feature that we could exploit for this purpose, perhaps a difference in the size of the bacterial and the Neanderthal DNA fragments? Alas, no! The size of bacterial DNA fragments in the bones was largely indistinguishable from that of the Neanderthal DNA.

Again and again we asked what differences there might be between bacterial and mammalian DNA. And then it struck me: methylation! Methyl groups are little chemical modifications that are common in bacterial DNA, particularly on A nucleotides. In the DNA of mammals, however, C nucleotides are methylated. Perhaps we could use antibodies to methylated A's to bind and remove bacterial DNA from the extracts. Antibodies are proteins that are produced by immune cells when they detect substances foreign to the body—for example, DNA from bacteria or viruses. The antibodies then circulate in the blood, bind with great strength to the foreign substances wherever they encounter them, and help eliminate them. Because of their ability to specifically bind to substances to which immune cells have been exposed, antibodies can be used as powerful tools in the laboratory. For example, if DNA containing methylated A nucleotides is injected into mice, their immune cells will recognize that the methylated A's are foreign and make antibodies to them. These antibodies can then be purified from the blood of the mice and used in the laboratory, and I thought we should make such antibodies and then try to use them to bind and eliminate bacterial DNA in our DNA extracts.

A quick literature search revealed that researchers at a company, New England Biolabs outside Boston, had already produced antibodies to methylated A's. I wrote to Tom Evans, an excellent scientist interested in DNA repair who I knew there, and he graciously sent us a supply. Now I wanted someone in the group to use them to bind to the bacterial DNA and remove it from the extracts. I thought that doing so would leave us with extracts in which the percentage of Neanderthal DNA was much higher. I considered this an ingenious plan. But when I presented it in our weekly meeting, people seemed skeptical—again, it seemed to me, because of their unfamiliarity with the technique. This time, bolstered by the fact that I had been right about the radioactivity, I more or less insisted. Adrian Briggs took it on. He spent months trying to get the antibodies to bind to the bacterial DNA and separate it from nonbacterial DNA. He tried all kinds of modifications of the technique. It never worked and we still don't know why. For quite some time, I got to hear facetious comments about my wonderful antibody idea.

What else could we try in order to eliminate the bacterial DNA? One idea was to identify sequence motifs found frequently among our bacterial

sequences. Perhaps we could then use synthetic DNA strands to specifically bind and remove the bacterial DNA in a way similar to what I had
imagined for the antibodies. Kay Pruefer, a soft-spoken computer science
student who, after coming to our lab, had taught himself more genome
biology than most biology students know, looked for potentially useful sequence motifs. He found that some combinations of just two to six nucleotides—such as CGCG, CCGG, CCCGGG, and so on—were present much
more often in the microbial DNA than in the Neanderthal DNA. When
he presented this observation in a meeting, it was immediately clear to me
what was going on. In fact, I should have thought of this earlier! Every molecular biology textbook will tell you that the nucleotide combination of
C followed by G is relatively infrequent in the genomes of mammals. The
reason is that methylation in mammals occurs to C nucleotides only when
they are followed by G nucleotides. Such methylated C's may be chemically modified and misread by DNA polymerases so that they mutate to
T's. As a result, over millions of years, mammalian genomes have slowly
but steadily been depleted of CG motifs. In bacteria, this methylation of C's
does not occur, or is rare, so CG motifs are more common.

How could we use this information? The answer to that question, too,
was immediately obvious to us. Bacteria make enzymes, so-called restriction enzymes, that cut within or nearby specific DNA sequence motifs
(such as CGCG or CCCGGG). If we incubated the Neanderthal libraries
with a collection of such enzymes, they would chop up many of the bacterial sequences so that they could not be sequenced but leave most of the
Neanderthal sequences intact. We would thus tip the ratio of Neanderthal
to bacterial DNA in our favor. Based on his analyses of the sequences, Kay
suggested cocktails of up to eight restriction enzymes that would be particularly effective. We immediately treated one of our libraries with this
mix of enzymes and sequenced it. Out of our sequencing machine came
about 20 percent Neanderthal DNA instead of 4 percent! This meant that
we needed only about seven hundred runs on the machines in Branford to
reach our goal—a number within the realm of possibility. This small trick
was what made the impossible possible. The only drawback was that the
enzyme treatment would cause us to lose some Neanderthal sequences—
the ones that carried particular runs of C's and G's—but we could pick up
those sequences by using different mixtures of enzymes in different runs
and by doing some runs without any enzymes. When we presented our restriction enzyme trick to Michael Egholm at 454, he called it brilliant. For
the first time, we knew that we could in principle reach our goal!

While all this was going on, a paper appeared by Jeffrey Wall, a young and talented population geneticist in San Francisco whom I had met on several occasions. It compared the 750,000 nucleotides that our group had determined by 454 sequencing from the Vi-33.16 bones and published in *Nature* with the 36,000 nucleotides that Eddy Rubin had determined by bacterial cloning from our extracts of the same bone and published in *Science*. Wall and his co-author, Sung Kim, pointed out several differences in the data sets, many of which we had already seen and discussed extensively when the two papers were in review. They suggested that there could be several possible problems with the 454 data set but favored the interpretation that there were huge amounts of present-day human contamination in our library. In particular, they suggested that between 70 and 80 percent of what we had thought was Neanderthal DNA was instead modern human DNA.[2]

This was troubling. We were aware that we might have some contamination in both the *Nature* and *Science* data sets, as we had sent the extracts to laboratories that did not work under clean-room conditions. We were also aware that if there was a difference in levels of contamination, they would be higher in the *Nature* data set produced at 454. We were sure, however, that any contamination levels could not be 70 to 80 percent, because Wall's analysis relied on assumptions, such as similar GC content in short and long fragments, that we knew were not true.

In an attempt to clarify these issues, we immediately asked *Nature* to publish a short note, in which we pointed out that several features differed between the sequences determined by the 454 technique and by bacterial cloning, and that some of the features were likely to affect the analysis. We also wanted to mention that our additional sequencing of the library had indicated very little contamination based on mtDNA. But we further realized that some level of contamination had probably been introduced into the library at 454, perhaps from a library of Jim Watson's DNA that it turned out 454 had sequenced at the same time as our Neanderthal library. So in the note we conceded that "contamination levels above that estimated by the mtDNA assay may be present." But by how much was impossible to tell. We pointed the readers both to Wall's paper and to the paper in which we described the use of tags in the library production that now made any contamination outside the clean room impossible.[3] We also posted a note in the publicly available DNA sequence database, so that any potential users would know of the concerns with these data. But, to my annoyance, after sending our note to reviewers, *Nature* decided not to publish it.

We discussed whether we had been too hasty in publishing the proof-of-principle data in *Nature*. Had we been driven to go ahead by the competition with Eddy? Should we have waited? Some in the group thought so, and others did not. Even in retrospect I felt that the only direct evidence for contamination we possessed, the analysis of the mtDNA, had shown that contamination was low. And that was still the case. The mtDNA analysis had its limitations, but in my opinion, direct evidence should always have precedence over indirect inferences. In the note that *Nature* never published, we therefore said that "no tests for contamination based on nuclear sequences are known, but in order to have reliable nuclear sequences from ancient DNA, these will have to be developed." This remained an ongoing theme in our Friday meetings for the next several months.

Chapter 14
Mapping the Genome

Once we knew that we could make the DNA libraries we needed, and with the hope that 454 would soon have fast enough machines to sequence them all, we started turning our attention to the next challenge: mapping. This was the process of matching the short Neanderthal DNA fragments to the human genome reference sequence. This process might sound easy, but in fact it would prove a monumental task, much like doing a giant jigsaw puzzle with many missing pieces, many damaged pieces, and lots and lots of extra pieces that would fit nowhere in the puzzle.

At heart, mapping required that we balance our responses to two different problems. On the one hand, if we required near-exact matches between the Neanderthal DNA fragments and the human genome, we might miss fragments that carried more than one or two real differences (or errors). This would make the Neanderthal genome look more similar to present-day humans than it really was. But if our match criteria were too permissive, we might end up allowing bacterial DNA fragments that carried spurious similarity to some parts of the human genome to be misattributed as Neanderthal DNA. This would make the Neanderthal genome look more different from present-day humans than in reality. Getting this balance right was the single most crucial step in the analysis because it would influence all later work that revolved around scoring differences from present-day genomes.

We also had to keep practical considerations in mind. The computer algorithms used for mapping could not take too many parameters into account, as it then became impossible to efficiently compare the more than 1 billion DNA fragments each composed of 30 to 70 nucleotides that we planned to sequence from the Neanderthal bones to the 3 billion nucleotides in the human genome.

The people who took on the monumental task of designing an algorithm to map the DNA fragments were Ed Green, Janet Kelso, and Udo

Stenzel. Janet had joined us to head a bioinformatics group in 2004 from University of the Western Cape, in her native South Africa. An unassuming but effective leader, she was able to form a cohesive team out of the quirky personalities that made up the bioinformatics group. One of these personalities was Udo, who had a misanthropic streak; convinced that most people, especially those higher up in academic hierarchies, were pompous fools, he had dropped out of university before finishing his degree in informatics. Nevertheless, he was probably more capable as a programmer and logical thinker than most of his teachers. I was happy that he found the Neanderthal project worthy of his attention even if his conviction that he always knew everything best could drive me mad at times. In fact, Udo would probably not have gotten along with me at all if it were not for Janet's mediating influence.

Ed, with his original project on RNA splicing having died a quiet, unmourned death, had become the de facto coordinator of the efforts to map the Neanderthal DNA fragments. He and Udo developed a mapping algorithm that took the patterns of errors in the Neanderthal DNA sequences into account. These patterns had in the meantime been worked out by Adrian together with Philip Johnson, a brilliant student in Monty Slatkin's group at Berkeley. They had found that errors were located primarily toward the ends of the DNA strands. This was because when a DNA molecule is broken, the two strands are often different lengths, leaving the longer strand dangling loose and vulnerable to chemical attack. Adrian's detailed analysis had also revealed that, contrary to our conclusions just a year earlier, the errors were all due to deamination of cytosine residues, not adenine residues. In fact, when a C occurred at the very end of a DNA strand, it had a 20 to 30 percent risk of appearing as a T in our sequences.

Ed's mapping algorithm cleverly implemented Adrian's and Philip's model of how errors occurred as position-dependent error probabilities. For example, if a Neanderthal molecule had a T at the end position and the human genome a C, this was counted as almost a perfect match, as deamination-induced C-to-T errors at the end positions of Neanderthal fragments were so common. In contrast, a C in the Neanderthal molecule and a T in the human genome at the end position was counted as a full mismatch. We were confident that Ed's algorithm would be a great advance in reducing false mapping of fragments and increasing correct ones.

Another problem was choosing a comparison genome to use for mapping the Neanderthal fragments. One of the goals of our research was to examine whether the Neanderthal genome sequence revealed a closer relationship with humans in Europe than in other parts of the world. For example, if we mapped the Neanderthal DNA fragments to a European genome (about half of the standard reference genome was from a person of European descent), then fragments that matched the European genome might be retained more often than fragments that were more like African genomes. This would make the Neanderthal incorrectly look more similar to Europeans than to Africans. We needed a neutral comparison, and we found one in the chimpanzee genome. The common ancestor that Neanderthals and modern humans shared with chimpanzees existed perhaps 4 million to 7 million years ago, meaning that the chimpanzee genome should be equally unlike both the Neanderthal and the modern human genome. We also mapped the Neanderthal DNA fragments to an imaginary genome that others had constructed by estimating what the genome of the common ancestor of humans and chimpanzees would have looked like. After being mapped to these more distant genomes, the Neanderthal fragments could then be compared to the corresponding DNA sequences in present-day genomes from different parts of the world and differences could be scored in a way that did not bias the results from the outset.

All of this required considerable computational power, and we were fortunate to have the unwavering support of the Max Planck Society as we attempted it. The society dedicated a cluster of 256 powerful computers at its computer facility in southern Germany exclusively to our project. Even with these computers at our disposal, mapping a single run from the sequencing machines took days. To map all our data would take months. The crucial task that Udo took on was how to more efficiently distribute the work to these computers. Since Udo was deeply convinced that no one could do it as well as he could, he wanted to do all of the work himself. I had to cultivate patience while awaiting his progress.

When Ed looked at the mapping of the first batches of new DNA sequences that came back to us from Branford, he discovered a worrying pattern that set off alarm bells in the group and made my heart sink: the shorter fragments showed more differences from the human genome than the longer fragments! It was reminiscent of one of the patterns that Graham Coop, Eddy Rubin, and Jeff Wall had seen in our *Nature* data. They had interpreted that pattern as contamination, assuming that the longer fragments showed fewer differences from present-day humans because many of them were in fact recent human DNA contaminating our libraries. We had hoped that preparing the libraries in our clean room and using our special TGAC tags would spare us the plague of contamination. Ed began frantic work to see if we after all had modern human contamination in our sequencing libraries.

Happily he found that we didn't. Ed quickly saw that if he made the cut-off criteria for a match more stringent, then the short and long fragments became just as different from the reference genome. Ed could show that whenever we (and Wall and the others) had used the cut-off values routinely used by genome scientists, short bacterial DNA fragments were mistakenly matched to the human reference genome. This made the short fragments look more different than long fragments from the reference genome; when he increased the quality cut-off, the problem went away, and I felt secretly justified in my distrust of contamination estimates based on comparisons between short and long fragments.

But soon after this, our alarm bells went off again. This time the issue was even more convoluted and took me quite a while to understand—so please bear with me. One consequence of normal human genetic variation is that a comparison of any two versions of the same human chromosome reveals roughly one sequence difference in every thousand nucleotides, those differences being the result of mutations in previous generations. So whenever two different nucleotides (or alleles as geneticists will say) occur at a certain position in a comparison of two chromosomes, we can ask which of the two is the older one (or the "ancestral allele") and which is the more recent one (or the "derived allele"). Fortunately, it is possible to figure this out easily by checking which nucleotide appears in the genomes of chimpanzees and other apes. That allele is the one that is likely to have been present in the common ancestor we shared with the apes, and it is therefore the ancestral one.

We were interested in seeing how often the Neanderthals carried recent, derived alleles that are also seen among present-day humans, as this would allow us to estimate when Neanderthal ancestors split from modern human ancestors. Essentially, more derived alleles shared by modern humans and Neanderthals means that the two lines diverged more recently. During the summer of 2007, Ed looked at our new data from 454 Life Sciences and he was alarmed. Just as observed by Wall and others in the smaller test data set published in 2006, Ed saw that the longer Neanderthal DNA fragments—those of more than 50 or so nucleotides—carried more derived alleles than shorter ones. This suggested that the longer fragments were more closely related to present-day human DNA than the shorter ones, a paradoxical finding that once again could have been the result of contamination.

Like many of the crises before it, this one dominated our Friday meetings. For weeks we discussed it endlessly, suggesting one possible explanation after another, none of which led us anywhere. In the end, I lost my patience and suggested that maybe we *did* have contamination, that maybe we should just give up and admit that we could not produce a reliable Neanderthal genome. I was at my wit's end, feeling like crying like a child. I did not, but I think many in the group realized it was a real crisis nonetheless. Perhaps this gave them new energy. I noticed that Ed looked as though he had not slept at all for a few weeks. Finally he was able to puzzle it out.

Recall that a derived allele starts out as a mutation in a single individual—a fact that, by definition, makes derived alleles rare. Examined in aggregate, one person's genome will show derived alleles at about 35 percent of the positions that vary whereas about 65 percent will carry ancestral alleles. Ed's breakthrough came when he realized that this meant that when a Neanderthal DNA fragment carried a derived allele, it would differ from the human genome reference sequence 65 percent of the time and match it only 35 percent of the time. This, in turn, meant that a Neanderthal DNA fragment was more likely to match the correct position if it carried the ancestral allele! He also realized that short fragments with a difference to the human genome would more often go unrecognized by the mapping programs than longer fragments, because the longer fragments naturally had many more matching positions that allowed them to be correctly mapped even if they carried a difference or two. As a result, shorter fragments with derived alleles would more often be thrown out by the mapping program than longer ones, and short fragments would therefore incorrectly seem to carry fewer derived alleles than longer fragments. Ed had to explain this to me many times before I understood it. Even so, I did not trust my intuition and hoped that he could prove to us in some direct way that his idea was correct.

I guess Ed did not want to see me cry in the meeting, so in the end he came up with a clever experiment that proved the point. He simply took the longer DNA fragments he had mapped and cut them in half in the computer, so that they were now half as long. He then mapped them again. Like magic, the frequency with which they carried derived alleles decreased when compared to the longer ones from which they were generated (see Figure 14.1). This was because many of the fragments that carried derived alleles could not be mapped when they were shorter. Finally, we had an explanation for the pattern of apparent contamination in our data! At least some of the patterns of contamination seen in the original test data published in *Nature* could also now be explained. I quietly let out a sigh of relief when Ed presented his experiment. We published these insights in a highly technical paper in 2009.[1]

Ed's findings reinforced my conviction that direct assays for contamination were necessary, and our Friday discussions again and again came back to how we could measure nuclear DNA contamination. But now I was somewhat more relaxed when these discussions came up. I felt convinced that we were on the right track.

Chapter 15
From Bones to Genome

By early 2008, the people at 454 Life Sciences in Connecticut had performed 147 runs from the nine libraries we had prepared from the Vi-33.16 bone, yielding 39 million sequence fragments. This was a lot but still not as much as I had hoped to have by this time, and certainly far too little to make it worthwhile to begin to reconstruct the nuclear genome. Nevertheless, I was keen to test the mapping algorithms, so we undertook the much less formidable task of reconstructing the mitochondrial genome. All we or anyone else had done by that point was sequence some 800 nucleotides of the most variable parts of Neanderthal mtDNA. Now we wanted to do all 16,500 nucleotides.

Ed Green began by sieving through the 39 million DNA fragments to identify those that were similar to the mtDNAs of present-day humans. He then compared these sequence fragments to one another to find where they overlapped, enabling him to build a preliminary Neanderthal mtDNA sequence. He next trolled through the 39 million sequences again, this time looking for the ones most similar to his preliminary Neanderthal mtDNA sequence that he might have missed the first time. In total, he identified 8,341 Neanderthal mtDNA sequences, averaging 69 nucleotides in length. From them he assembled a complete mtDNA molecule of 16,565 nucleotides, the longest contiguous Neanderthal mtDNA sequence ever reconstructed.

This gave me a comforting sense of having achieved something concrete, although the analysis of the Neanderthal mtDNA genome revealed nothing about Neanderthals that we had not already known. The useful insights that we did find were of a technical nature. For example, we found that the number of fragments we retrieved varied across the genome. Ed realized that this correlated with the amounts of G and C nucleotides relative to A's and T's in the fragments. This meant that DNA molecules rich in G and C survived better in the bone—or, perhaps, survived our extraction of DNA from the bone better. But the good news was that no parts of the mtDNA were missing. I began to feel that many of the technical problems in analyzing Neanderthal DNA fragments were now under control. We also found 133 positions where the Neanderthal mtDNA differed from all, or almost all, human mtDNAs today.[1] Before this, we had known of only three such positions in the short segment of Neanderthal mtDNA we had published in 1997. Using the 133 positions, we could now more confidently estimate the level of modern mtDNA contamination in our new data. It was 0.5 percent. We also went back and estimated the mtDNA contamination in the old test data in our 2006 *Nature* paper as well as the additional data generated while our *Nature* paper was still in press. Out of 75 mtDNA fragments, 67 were of the Neanderthal type. So there was 11 percent contamination in that library, more than we had hoped but much less than the 70 to 80 percent suggested in the Kim and Wall paper. We included all this information in a paper we submitted to *Cell*, the journal that published our initial Neanderthal mtDNA results in 1997. Again we stressed that a direct way to estimate contamination specifically for the nuclear genome would be better. And we renewed our discussions in the Friday meetings about how best to do this.

Once that analysis was finished, however, I became worried that the mtDNA paper had diverted our attention from the fact that the accumulation of Neanderthal DNA sequences was going slowly. We were now well into the second year of the project, and only months away from the two-year deadline we had publicly set ourselves for producing the 3 billion Neanderthal nucleotides. Being a bit overtime would be a small thing, but unfortunately I felt we were on track to make a much worse finish. As a result, lab meetings were becoming increasingly tense. I sometimes found myself becoming loud and sarcastic (which I sorely regretted later), generally because of some illogical arguments or because someone was unable to succinctly describe what he or she had done in the lab. But the deeper reason for my bad temper was my perception that the project was not moving forward. Part of the reason for the slow progress was that few extracts contained enough Neanderthal DNA to enable production of DNA libraries at the speed we had hoped, but it was also obvious that the sequencing at 454 was not going quickly. Michael Egholm certainly remained committed to the project, but 454 Life Sciences had been sold to the Swiss pharmaceutical giant Roche in March 2007. As a result, the person who handled the day-to-day sequencing at Branford left the company the following fall, and I suspected that it had become hard for Egholm and the others to devote their full attention to the Neanderthal genome. For the first time, I flirted with the idea of working with 454's competition.

One of those competitors was David Bentley, an accomplished human geneticist I met at the Cold Spring Harbor meeting in May 2007. In 2005 he had moved from the Wellcome Trust Sanger Institute to Solexa, a new company spun out of the chemistry department of the University of Cambridge. At Solexa he oversaw the development of a DNA sequencing machine that presented the strongest competition yet for Jonathan Rothberg's 454. Just like the 454 technique, the Solexa technique used adaptors stuck to the ends of molecules to create DNA libraries that could be amplified and sequenced. Unlike the 454 technique, however, each library molecule was amplified not in little oil droplets but by primers attached to a glass surface. So from each initial DNA strand that had landed on the glass, a little spot, or cluster, composed of millions of copies of the original library molecule would emerge. These clusters were then sequenced by the addition of a sequencing primer, a DNA polymerase, and the four bases, each labeled with a different fluorescent dye.

The first test versions of these machines were delivered to sequencing centers in 2006. They could sequence a stretch of only twenty-five nucleotides, and I had heard at the time that the machines tended to break down a lot. But the great potential advantage of this technology was that each run of the machine allowed the sequencing of not hundreds of thousands of individual DNA fragments, as on the 454 machines, but a few million, and this number could potentially increase as the machines improved. Soon, too, the read lengths became 30 nucleotides, and there was talk about an improvement that would allow each DNA fragment to be sequenced from each of its ends so that a total of 60 nucleotides could be read. This began to sound very interesting indeed for ancient DNA researchers. Others were also interested. In November 2006, Illumina, a US-based biotech company, bought Solexa. David Bentley was now the new company's chief scientist and vice president.

At the Cold Spring Harbor meeting, I discussed our project with David. He agreed that I could send him a mammoth or Neanderthal extract to test how Illumina's technology worked. In fact, we had already begun such a test. We'd been so eager to try this technology that a few months earlier, in February 2007, we sent one of our best mammoth DNA extracts to Jane Rogers at the Sanger Institute in Cambridge, who was in charge of their Solexa machines. We had not heard back from her yet, however, so I returned from Cold Spring Harbor with a new sense of urgency and began pestering our contacts at Sanger about the results. In early June, the data came back. We were a bit disappointed to see from the sequences that the technology seemed error prone. The company worked hard on improving this, but I also realized that the error rate could be compensated for by the very large numbers of DNA fragments the machines could sequence. In principle, we could simply sequence every DNA fragment in a library multiple times so that the errors would be easy to spot and disregard. Unfortunately, Illumina did not operate its own sequencing center, as did 454, so we would need our own machine, which, due to high demand, we only got six months later. By now it could read 70 nucleotides, but it still made many errors, which became more and more frequent further along into the sequences it read. A technical upgrade of the machine in 2008 allowed us to sequence each DNA fragment in our libraries from each of its ends. Since our Neanderthal DNA fragments were on average just 55 nucleotides long, we could therefore read each DNA sequence twice, once from each side. This meant that for most of each fragment, we had reliable sequence information.

The person who took on the challenge of analyzing the Illumina data was Martin Kircher, who had joined Janet Kelso's bioinformatics group as a graduate student in the summer of 2007. His boyish looks and charming smile belied what I felt was an overconfidence in himself that bordered on arrogance, perhaps inspired by his unofficial mentor Udo. This had greatly irritated me initially, but gradually I came to realize that his opinions were actually often right. I learned to appreciate his ability to quickly grasp technical issues, organize the flow of data from the sequencing machine into the cluster of computers, and give feedback to the technicians operating the machines. And he worked incredibly hard. Not only I but also Janet and everybody else came gradually to rely more and more on Martin for keeping our Illumina sequencing machine going and pushing analyses through the computers.

By early 2008 it was clear that we needed to abandon the 454 technology altogether to have a chance to finish the Neanderthal genome in a reasonable amount of time. The strong point of the 454 technology was its ability to sequence long DNA fragments, but since our DNA fragments were short, this was of no interest to us. Our goal was to sequence many short DNA fragments as quickly as possible. And in such mass production, Illumina had a real advantage over 454. But moving on from the 454 approach would not be straightforward, as Ed Green and the others had been busily building programs designed to handle 454 data. Switching to

Illumina would mean revamping the data-handling procedures and merging the different versions of sequence data. The technologies were so new that off-the-shelf software to solve such problems was not available. We had to do it all ourselves.

As the summer of 2008 approached, these issues came to a head. In mid-July it would be two years since our press conference. Clearly, we would not make that deadline, but if journalists called and asked, I wanted at least to have a new timeline. We now had enough bones and DNA extracts to make sufficient sequencing libraries to arrive at 3 billion nucleotides, but the only way to sequence the entire genome as we had promised was to move to Illumina. Finally, I took a substantial amount of the money we had set aside to pay 454 for sequencing and instead ordered four more Illumina machines. With five machines working simultaneously, I thought we could do it, and if the machines were delivered promptly, we might even be able to do it by the end of the year. Again, I had to end a collaboration, and toward this end I met Michael Egholm from 454 at a meeting in Denmark. Fortunately, he understood my reasoning, but he predicted that we would regret having to deal with "error-ridden microreads," as he disparagingly termed the shorter reads produced by Illumina machines.

In the middle of all these emotional ups and downs, I was happy to take a private respite from the Neanderthal work. On July 1, Linda and I flew off to Kona, on Hawaii. The official reason for the trip (and the explanation I gave to the lab) was that I had been invited to something called the Academy of Achievement, an annual conference at which musicians, politicians, scientists, and authors come together in a secluded and intimate environment to share ideas and experiences with one another and with a hundred graduate students from all over the world. Excited as I was to spend a few days with many famous and wise people, this was not the main reason I had come here. Linda and I had decided to use this opportunity to get married. It was something we had long postponed, mainly because I considered marriage an old-fashioned formality. We had decided to marry now in part because of practical reasons having to do with German pension plans in case I would predecease her, but we wanted to do it in private and with a slightly antic touch. We arranged a ceremony with a New Age pastor on the beach, a setting as beautiful as one might imagine. The pastor began by invoking Hawaiian spirits, blowing long tones on a conch shell to the four cardinal points of the compass. We made our private vows to each other and she pronounced us man and wife. In spite of the practical considerations that had prompted our decision to marry, I felt the ceremony manifested a deep commitment that had developed between us. My life with Linda had proved to be much richer than had my monk-like existence as a professor in Munich, particularly with our son Rune's birth in 2005.

After the ceremony on the beach, we took off on a hike. Linda had found an area in a park on the Big Island that was both lonely and beautiful. We started out walking with our heavy backpacks in the glaring sun over moon-like lava beds until we reached the coast. There we spent four days strolling naked on pristine beaches, snorkeling with fishes and turtles, and making love on the beach and under the palm trees. When I fell asleep to the soothing swell of the Pacific Ocean and the rustling of the palm leaves overhead, the subarctic steppe where the Neanderthals had lived seemed very far away. It was the perfect interruption in an extremely intense phase of my life.

But the Hawaiian interlude was short. The week after Linda and I came back from Hawaii, I gave a talk at the World Congress of Genetics, in Berlin. I described our technical progress toward sequencing the genome and our mtDNA results. It was frustrating to not have more to say. Another speaker at the meeting was Eric Lander, one of the main thinkers and driving forces behind the public effort to sequence the human genome. I tended to find him almost intimidatingly insightful and sharp. I had often encountered Eric at Cold Spring Harbor and in Boston, where he headed the Broad Institute, a very successful research institute devoted to genomics, and I had frequently profited from his advice. After the congress, he came to Leipzig to visit our group. We had still not taken delivery of our four new Illumina machines, and with our single machine we could not generate data fast enough, since each run took two weeks plus processing time on the computers. Fortunately, Eric, a great champion of the Illumina machines, had several at the Broad Institute, and he offered to help. We were only days from our self-imposed deadline, so I took him up on his offer without hesitation. We would obviously not be done by our deadline, but if we generated the data by the end of 2008, we would at least have done it within the year we said we would.

As our two-year deadline came around, both *Nature* and *Science* started courting us to submit our Neanderthal genome paper to them. I considered doing what I had done in 1996 with the first Neanderthal mtDNA sequences-publishing in *Cell*, which is a more serious molecular biology journal. But there was something to be said for publishing in either of the two other prestigious journals, where everybody expected to see the work—especially the students and postdocs, who tended to think that their careers might benefit from publishing there. In June, Laura Zahn, an editor from *Science*, visited us to discuss the Neanderthal paper. *Science* is published by the American Association for the Advancement of Science, and shortly after Laura's visit the AAAS invited me to give a plenary presentation on the Neanderthal work at its annual meeting, which would take place in Chicago on February 12–16, 2009. This provided a definitive deadline to work against, one I felt certain we would make. So I agreed to the talk, and I realized that this meant we would most likely publish in *Science*.

As so often happens, things took longer than expected. It took us until the end of October to produce five Illumina-style DNA libraries with our special Neanderthal tags. We sequenced aliquots of each library on our Illumina machine and carefully determined the numbers of molecules in the libraries. We found that they contained over 1 billion DNA fragments. This should give us what we needed to complete the genome sequence. We sent the libraries along with custom primers to the Broad Institute for sequencing. The output from the Illumina machines, however, we would analyze using a computer program developed by Martin Kircher that read more nucleotides with fewer errors than the commercial programs provided by Illumina's program could. The amount of data from the sequencing machines that his program needed was so huge that it was impractical to transfer it via the Internet; instead, we had arranged for large-capacity computer hard drives to be shipped from Boston to Leipzig.

In mid-January 2009, we took delivery of two hard drives containing the results of the first few runs. Now the special tags on our Neanderthal libraries proved their usefulness. Martin discovered that one of the runs from the Broad Institute contained *no* sequences with these tags. Clearly, something had gotten mixed up at the Broad Institute. This was startling, and a bit alarming. I considered shifting the sequencing back to our lab, where by now we had our four new Illumina machines up and running. But the other runs on the two hard drives from the Broad Institute looked good, and we were committed to working with Eric. Finally, on February 6, 2009, eighteen hard drives arrived by FedEx. It was not a day too early. In six days, on February 12, the AAAS meeting would convene.

Martin, Ed, and Udo checked the data from the Broad Institute. The reads carried our tags, the distribution of fragment sizes were the same as what we had seen in our own Illumina runs, and when Udo mapped the reads the results were consistent with the data we had generated in Leipzig. This was a relief. The AAAS had pushed for a press conference to accompany my talk at its Chicago meeting, and I had dreaded having nothing to say. Now I would be able to announce that we had produced the sequences needed to arrive at 1-fold coverage of the genome. But just as I had wanted the initial announcement of the project to be in Leipzig, a city still in the process of emerging from the shadows of its socialist past, I now felt that the AAAS press conference should take place in Leipzig, too. And in recognition of the early support from 454 Life Sciences for our project, I wanted to organize the press conference with them. The AAAS agreed and for February 12 we organized a press conference with 454 in Leipzig and with a video link to Chicago, so that the meeting participants and the Chicago press could ask questions. I would then fly to Chicago for my talk, scheduled for February 15.

This left us just six days to prepare. I focused the largest part of the press release, and my talk in Chicago, on the technical obstacles we had had to overcome in order to arrive at our first view of the genome of an extinct human form. I described how Tomi Maricic had used minute amounts of radioactively labeled DNA to identify and modify the steps where losses of DNA occur, how the tagged libraries produced in our clean room eliminated the problems of contamination that had affected the pilot study, how the detailed studies by Adrian Briggs and Philip Johnson had revealed patterns of errors in the DNA sequences, and how the computer programs Udo Stenzel and Ed Green had developed allowed us to identify and map the Neanderthal DNA fragments while avoiding many pitfalls.

I also wanted to say something about Neanderthals. We had not had time to map, much less analyze, the billion or so DNA sequences. Fortunately, over the past six months, Udo and others had mapped the more than 100 million DNA fragments we had sequenced with the 454 technology. This allowed us tease out a few tidbits of biological relevance. Ed had looked at two cases where others had claimed that gene variants now seen in present-day humans were likely to have been contributed by gene flow from Neanderthals. One of these was a big region of 900,000 bases on chromosome 17 that is inverted (or reversed) on the chromosome in many Europeans. The excellent Icelandic genealogical records had been used to show that the inverted form was associated with slightly higher fertility in women. Did the inverted version come from Neanderthals, as some people had speculated? Ed checked our Neanderthal sequences, and none of the three Neanderthals who had contributed sequences to our effort carried the inverted version. This finding did not rule out the possibility that other Neanderthals may have carried the inverted variant and contributed it to Europeans, but it made it less likely. Similarly, a gene on chromosome 8 that, when mutated, drastically reduces brain size comes in different versions in normal people around the world; the version common in Europe and Asia was suggested to have come from Neanderthals. But Ed showed that our sequences did not carry that variant. So, from looking at these examples, there was no hint of a genetic contribution from Neanderthals to modern Europeans. I was comfortable with this conclusion, which fit what we had found a decade before with the mitochondrial DNA data. But I was to be startled by the results of some other last-minute discoveries.

Chapter 16
Gene Flow?

During the long flight from Chicago back to Leipzig, I tried to soberly assess the status of the project. Although we had now generated all the DNA sequences we needed, much work remained. The first thing we needed to do was to map all the DNA fragments sequenced with the Illumina technology to the chimpanzee genome and to the reconstructed ancestral genome of humans and chimpanzees. The group in Leipzig would now focus on adapting the algorithms that Ed and Udo had developed for the 454 data to our new Illumina data.

Once this was done, we could start asking several questions about our relationship with Neanderthals: when our lines diverged, how different we were, whether our lines had ever mixed, and whether any genes had changed in interesting ways between people today and Neanderthals. To answer such questions, we would need more than just the people in my group—we would need many people from all over the world.

Back in 2006 I had come to realize that our project was historic not only because this was the first time the genome of an extinct form of human was to be sequenced, but also because it was the first time that a small academic group had taken on the sequencing of an entire mammalian genome. Until that time, only large sequencing centers would have been able to undertake such a project. But even those large centers collaborated with other institutions to analyze different aspects of the genomes. We clearly needed to put together some sort of consortium. So in 2006, I started thinking about what types of expertise we would need and which people I would like to work with.

First and foremost, we needed population geneticists. These are geneticists who study DNA sequence variation in a species or a population and, from this, infer what happened to these species or populations in the past. They can tell when populations split, whether they exchanged genes, and whether selection acted upon them. The population geneticists in our group, Michael Lachmann and Susan Ptak, could help with some of these things but we clearly needed input from more people, and we wanted to work only with the very best.

I started contacting people, most of them in the United States, as soon as our project was under way. Almost everyone I talked to wanted to be part of the project—it was clearly a unique opportunity to study a genome most researchers had thought impossible to sequence—but we needed people willing to work full-time or almost full-time on the project for at least a few months so that we could finish the analyses quickly. I had seen too many examples of genome projects that dragged on for months or years because crucial groups had multiple and conflicting commitments. To their credit, when I made this clear, several people realized they had too many other things to do and backed out.

One person I particularly wanted in the group was David Reich, a young professor at Harvard Medical School and a rather unorthodox population geneticist. He first studied physics at Harvard and then went on to do a PhD in genetics at Oxford. I invited him to visit Leipzig in September 2006, and he gave a talk on a controversial paper he and his colleagues had just published that summer in *Nature*.[1] It suggested that after the initial separation of the populations that would become humans and chimpanzees, the two populations came together more than a million years later and exchanged genes before separating permanently. I found David to be very stimulating to talk to. In fact, I found him on the verge of intellectually intimidating. He produced a torrent of thoughts and ideas at a rate that was challenging and, at times, almost impossible to keep up with. But the intellectual onslaught was balanced by the fact that David is the kindest, gentlest person imaginable. He was and is also remarkably unconcerned about academic prestige. He shares, I imagine, my conviction that academic positions and grants will be available if one simply does good work on interesting problems. I spoke with him about the Neanderthal project during his visit to Leipzig and gave him our manuscript on the pilot study to read on his flight back to Boston. A few days later I received six pages of detailed comments on our paper. It was clear that he was an ideal candidate to work with on the Neanderthal genome.

In fact, working with David would mean that we would have not only his amazing brain involved in the project but also the unique capabilities of his close associate Nick Patterson. Nick had had an even more unusual career than David. He had studied mathematics at Cambridge in the UK and then worked for the British intelligence service as a cryptologist for over twenty years. Some people I have since met have said that at that time he had the reputation of being one of the best code breakers in both the British and US intelligence communities. After leaving the secretive intelligence world, he turned his attention to predicting financial markets; by 2000 he had earned enough money on Wall Street to live comfortably for the rest of his life. Ever intellectually curious, he then moved on to what would later become the Broad Institute in Boston to use his code-cracking abilities on the deluge of genome sequences that were generated there. In Boston, he eventually joined forces with David. Nick is the epitome of what a child might imagine a brilliant scientist to look like. Due to a congenital bone disease his head seems disproportionately large and his eyes are directed in different directions. This makes him seem constantly concerned with higher mathematical problems. I came to learn that he was also a Buddhist, sharing my long-standing but unfortunately not very committed interest in Zen Buddhism. Nick has an uncanny ability to discern patterns hidden in large amounts of data. I was so excited by the prospect of having Nick and David involved in our project that I offered to hire them both for the duration of the project if they would spend at least 75 percent of their time in Leipzig. Although they couldn't accept this offer, they promised to devote as much attention as they could to the Neanderthal genome, a promise that they would fulfill to an extent that exceeded even my greatest expectations.

Another population geneticist I wanted on board was Montgomery, or "Monty," Slatkin. He was based at UC Berkeley, where I had first met him in the 1980s when I was a postdoc with Allan Wilson. Monty has had a long and distinguished career as a mathematical biologist and has the level-headedness and balance that are the hallmarks of wisdom and experience. He had trained many brilliant students who went on to head their own groups, and the younger people who worked with him then were equally promising. Foremost perhaps was Philip Johnson, who was to work out the patterns of errors in the Neanderthal sequences alongside Adrian Briggs (see Chapter 14). I was delighted that Monty wanted to join our consortium, not least because his scientific style balanced David's and Nick's. Whereas they liked to come up with clever algorithms to infer past population events, Monty liked to construct explicit population models and test whether they fit the variation observed in DNA sequences.

One of the first questions the consortium wanted to take on was perhaps the most hotly contested one: Had Neanderthals contributed DNA to people living in Europe today? After all, they had lived throughout Europe until modern humans appeared around 40,000 years ago, and some paleontologists claimed that they saw Neanderthal traits in the skeletons of early modern people in Europe. The majority of paleontologists disagreed, however, and our 1997 paper analyzing Neanderthal mtDNA had given no hint that they had contributed DNA to present-day Europeans. Only an analysis of the nuclear genome would be able to definitively answer the question.

To understand why an analysis of the nuclear genome would be so much more powerful than analysis of the mtDNA, it's important to remember that whereas the nuclear genome is composed of over 3 billion nucleotides, the mtDNA genome is made up of only 16,500 nucleotides. In addition, the nuclear genome is reshuffled in each generation, when each chromosome in a pair exchanges pieces with its partner and each chromosome is passed on independently of the others to offspring. Due to the shuffling as well as the sheer size of the nuclear genome, there are many chances to see even a small amount of mixing between two groups. If a child were born from a union between a Neanderthal and a modern human, then it would get roughly 50 percent of its DNA from each of the two groups. If that child then grew up with modern humans and in turn reared children with them, its children would carry, on average, 25 percent Neanderthal DNA, its grandchildren would carry 12.5 percent, its great-grandchildren would carry about 6 percent, and so on. Although the contribution decreases rapidly in this scenario, 6 percent of the genome still represents over 100 million nucleotides. And eventually the Neanderthal DNA would spread throughout the population so everyone would have some proportion of it. At that point, when both parents of a child would carry roughly similar proportions of Neanderthal DNA, it would not become further diluted but remain indefinitely in the population. Also, if mixing did happen, it likely didn't occur only once. And if the population where the mixed children lived ended up expanding so that, on average, there would be more than one child per person in the next generation, then the contribution would tend to not get lost. We of course know that the modern human population expanded after they came to Europe and replaced the Neanderthals, so I felt certain that we would be able to see even a rather small contribution if it existed. But since the mtDNA showed no signs of a contribution, I still tended to think that no contribution had occurred.

I was skeptical of a Neanderthal contribution in part because I suspected that some biological problem might interfere with successful mating. Although Neanderthals and modern humans almost certainly had sex with each other—after all, what humans groups don't?—I sometimes wondered if perhaps some factor might have made the offspring less fertile. For example, humans have 23 pairs of chromosomes whereas chimpanzees and gorillas have 24 pairs. This is because one of our largest chromosomes, chromosome 2, is a fusion of two smaller chromosomes that still exist in the apes. Such rearrangements of chromosomes occasionally occur during evolution and generally are of no consequence for how the genome functions. But the hybrid offspring of individuals that have different numbers of chromosomes often have difficulty conceiving offspring themselves. If the fusion that created chromosome 2 occurred after modern humans diverged from Neanderthals, then perhaps we interbred with them but the children didn't transmit any Neanderthal DNA because they couldn't have children of their own. But these were just idle musings; now we hoped to actually find out for certain. And the best way to do this was to compare the Neanderthal genome to the genomes of present-day people to see if it was closer to people in Europe, where Neanderthals had lived, than to people in Africa, where they had never lived.

By October 2006, David and Nick were already deeply immersed in the project. They worked with Jim Mullikin, another member of our consortium. Jim was the head of DNA sequencing at the National Human Genome Research Institute (NHGRI) in Bethesda. He was soft-spoken and immensely helpful. In fact, he somehow reminded me of Winnie the Pooh, but a very, very competent version of the friendly bear. Jim had sequenced genomes from several modern Europeans and Africans. To compare these genomes with the Neanderthal genome, he identified positions where a nucleotide in one individual in such a pair differed from the other individual. Such positions, which, as noted, are called single nucleotide polymorphisms or SNPs, are the basis for almost all genetic analyses. I remembered how excited I was back in 1999 when Alex Greenwood found the first Ice Age SNP ever seen (see Chapter 9); he had retrieved nuclear DNA sequences from a mammoth and found a position where the two mammoth chromosomes differed from each other. Now we wanted to analyze the hundreds of thousands of such SNPs that had been identified in humans to see which versions the Neanderthals had carried about 40,000 years ago, long before the Ice Age when the mammoth had lived. Although we had worked toward this goal for many years, it still seemed like science fiction to me.

To use SNPs to look for traces of possible interbreeding between Neanderthals and modern humans, we went back to the logic underpinning an analysis we had done of the first Neanderthal mtDNA in 1996. At that time, we argued that since Neanderthals were known to have lived only in Europe and western Asia, we would expect any contribution of their mtDNA to have occurred there. Thus, if Neanderthals and modern humans had mixed, some Europeans would walk around with mtDNA that until some 30,000 years ago had been in a Neanderthal. So we would expect Neanderthal mtDNA to be on average more similar to that found in some Europeans than to that found in people in Africa. We had failed to see this and thus concluded that no mtDNA contribution had occurred. In the case of the nuclear genome, the same argument would hold; if there were no contribution from Neanderthals to present-day humans anywhere in the world, then on average, across many individuals and many SNPs in the genome, the Neanderthals should carry equally many nucleotide differences from all populations. If, on the other hand, they had contributed to some population, genomes in that population would on average be closer to the Neanderthal genome than to genomes in other populations. David, Nick, and Jim would therefore identify SNPs where one of the Africans Jim had sequenced happened to differ from one of the Europeans. They would then count how often the Neanderthal genome matched the African and European genomes, respectively. If Neanderthals were closer to the Europeans, that would indicate gene flow from Neanderthals into European ancestors.

In April 2007, in preparation for the Cold Spring Harbor Genome Meeting, Jim and David sent me their first analysis of the Neanderthal sequences we had generated with the 454 technique. To test the method, they had first analyzed a present-day European individual at SNPs where another European and an African were known to differ from each other. They found that the European matched the other European at 62 percent of the SNPs and the African at 38 percent of the SNPs. Thus, as we had expected, on average people from the same part of the world shared more SNP variants with each other than with people from other parts of the world. They were able to compare the Neanderthal sequence to 269 positions where the European and African individuals differed and they found that the Neanderthal matched the European at 134 positions and the African at 135 positions. This was as close to 50:50 as the data could possibly be and it perfectly fit my preconceived idea that there had been no admixture. I liked this result for another reason as well. It meant that what we had was the DNA of a person who seemed to be equally related to Europeans and Africans. In short, there couldn't be very much DNA contamination from present-day humans among our Neanderthal sequences, since any such contamination would likely have come from a European individual and therefore make the Neanderthal look closer to the European than to the African individual.

On May 8, 2007, the day before the meeting started, all the members of what was now officially called the Neanderthal Genome Analysis Consortium met for the first time at Cold Spring Harbor. I started the meeting by describing the tag that we had introduced to rule out any contamination that occurred after the libraries left our clean room. I also talked about the three archaeological sites (see Chapter 12) and the bones from which we now had generated data. We had 1.2 million nucleotides of Neanderthal DNA determined from Vindija with our new tagged library approach. We also had about 400,000 nucleotides from the type specimen from Neander Valley in Germany, the bone from which we had determined the mtDNA segment in 1997. Finally, we had 300,000 nucleotides from El Sidrón, the cave in Spain where Javier Fortea and his team had collected bones under sterile conditions for us.

My description of the Neanderthal sites was a welcome relief from the rather arcane technical discussion of how we had extracted and sequenced DNA from the bones and how these sequences might be analyzed. Everybody was impressed that the Neanderthal seemed to be equally distant from an African and a European individual, but David Reich correctly pointed out that with just 269 SNPs, we could only exclude a very large genetic contribution from Neanderthals into Europeans. In fact, the 90 percent confidence interval for the 49.8 percent estimate of the SNPs matching the European was 45.0 to 55.0 percent. This meant that, with 90 percent confidence to be correct, we could only say that Neanderthals hadn't contributed more than 5 percent of the genome to Europeans. In other words, there was a 10 percent chance that Neanderthals had contributed more than 5 percent. This uncertainty drove home for me a powerful advantage of molecular genetic analyses over paleontological analyses. If we had been discussing the forms, shapes, holes, and ridges of Neanderthal bones, we couldn't have made any realistic estimate of how sure we were of what we found. Neither could we have been confident of being able to collect more data to resolve the issue with greater confidence. With DNA, we could.

David had also used the SNPs Jim had detected in present-day humans in other analyses. He compared the DNA sequence for each SNP to that of the chimpanzee to determine which of the two variants, or alleles, was ancestral, and which was derived. The further back in time the Neanderthal population became separated from modern human populations, the less often the Neanderthal would carry the newer, derived SNP alleles found in people today. When David analyzed 951 SNPs that had been discovered in Africans, he found that a present-day European carried derived alleles at 31.9 percent of the SNPs. When he analyzed our Neanderthal sequences, he found that they carried the derived alleles at 17.1 percent of the SNPs, about half as often as the present-day European. Given certain assumptions, such as constant population size over time, this suggested that the Neanderthals separated from Africans some 300,000 years ago. I was delighted by these results. The sequences we had determined clearly came from a creature with a history very different from that of people living today. However, David dampened my enthusiasm by again pointing out that we didn't have very much data yet. In fact, the 90 percent confidence interval for the percentage of derived alleles in the Neanderthal ranged from 11 to 26 percent. Still, we were clearly on the right track.

After we shifted to using the Illumina sequencing machines and began generating DNA sequences at a much faster rate, our twice-per-month phone meetings with the consortium became longer and we started having them every week. In January 2009, as the AAAS meeting drew near, I pleaded with David and Nick to do a quick analysis of our 454 sequences, which represented about 20 percent of all our data. Although I still didn't think there had been any interbreeding between Neanderthals and modern humans, I wanted David to come up with an estimate of how small any contribution by Neanderthals to Europeans could maximally be without our detecting it. In other words, how big a contribution could we exclude? That was the number I wanted to present at the press conference and the AAAS meeting.

On February 6, 2009, I received an e-mail from David. It said, "We now have strong evidence that the Neanderthal genome sequence is more closely related to non-Africans than to Africans." I was totally taken aback. David found that our Neanderthal sequences matched Europeans at 51.3 percent of SNPs. This may not seem very different from 50 percent, but we now had so much data that the uncertainty was just 0.22 percent, which meant that even if we subtracted 0.22 from 51.3, we still had a number that was different from 50 percent. I realized I might have to revise my ideas and concede that there had been genetic mixing between Neanderthals and the ancestors of Europeans. But there was another observation that made me wonder if there was something wrong with the analysis after all. When David compared the Chinese and African genomes, the Neanderthals matched the Chinese 51.54 percent of the time and the uncertainty was 0.28 percent despite the fact that there had never been Neanderthals in China. David himself was intrigued as well as worried by these results. We both agreed that this finding was potentially very exciting but also that the results had the potential to be spectacularly wrong. There was a frantic exchange of e-mails and David, Nick, and I agreed that we should keep our admixture result secret at the press conference and AAAS meeting. If we mentioned it, all the media would write about it. If it then later turned out to be due to some sort of error, we would look like idiots. Instead, I decided to talk about less hot topics in Chicago. Discussions about the potential admixture would have to be postponed to a meeting that the consortium would have in Croatia just after the AAAS meeting.

Chapter 17
First Insights

Two days after returning from Chicago, I was again in an airplane, this time on my way to Zagreb to give a lecture on our project at the Croatian Academy of Sciences and Arts. The next day I flew south to Dubrovnik, where our consortium and our Croatian collaborators were to meet in a hotel on the coast outside the city. We were not there just to celebrate, but to hammer out how we would analyze and publish the Neanderthal genome.

But the flight to Dubrovnik didn't go quite as planned. The Dubrovnik airport is squeezed between mountains and the sea and has a bad reputation for difficult side winds. It was at this airport that US Secretary of Commerce Ron Brown died in an airplane accident in 1996. The US Air Force investigation later attributed the crash to pilot error and a poorly designed landing approach. As we approached the airport it was windy and the plane was jumping around. The Croatian pilot, probably wisely, decided not to try to land. Instead he flew to Split, a city some 230 kilometers away. We arrived there late in the evening and were packed into an overcrowded bus that took us through the night to Dubrovnik. I was exhausted when our first session started at 9:00 a.m.

Despite how tired I was, I felt energized by the presence of almost all the twenty-five members of our analysis consortium in the conference room (see Figure 17.1). Together, we were now going to tease out the information in the 40,000-year-old DNA sequences we had determined. I gave the first talk, an overview of the data we now had in hand. This was followed by a technical presentation by Tomi about his library preparation. Ed described how we estimated the level of present-day human DNA contamination, the issue that had plagued our first paper back in 2006. Our "traditional" mtDNA analysis yielded an estimate of 0.3 percent. By the time of the meeting, we had also devised an additional analysis not based on mtDNA. It relied on using the large numbers of DNA fragments we had from certain portions of the genome—specifically, the sex chromosomes,

FIGURE 17.1. The consortium meeting in Dubrovnik, Croatia, in February 2009. Photo: S. Pääbo, MPI-EVA.

X and Y. Because females carry two X chromosomes while males carry one X chromosome and one Y chromosome, if a bone came from a female, we should find only X chromosome fragments and no Y chromosome fragments. Therefore, any Y chromosome fragments we detected in libraries derived from a female's bone would be indicative of contamination by modern males.

This analysis, suggested during one of our Friday meetings in Leipzig, initially sounded simple. But as with so many of the things Ed did, it was not as straightforward as it seemed. The complication was that, although the X and Y chromosomes are morphologically distinct, some of their parts share a close evolutionary relationship. The DNA they share as a result of this relationship could confuse the analysis when we mapped our short DNA fragments. To avoid this problem, Ed identified 111,132 nucleotides on the Y chromosome that weren't similar to anything else in the genome, even if these bits were fragmented into pieces as small as 30 nucleotides in length. When he looked among the Neanderthal DNA fragments, he found just four fragments carrying these Y chromosomal sequences; if the bones we used had all come from males, he would have expected to see 666 of them.

Thus, he inferred that all three bones came from female Neanderthals, and that the four Y chromosomal DNA fragments must have come from DNA contamination. This suggested that we had about 0.6 percent male contamination. This estimate was not perfect since we could detect only contaminating DNA from men, but it suggested that the level of contamination was low and similar to what we had estimated from the mtDNA.

We discussed other ways to estimate contamination. Philip Johnson from Monty's group at Berkeley suggested an approach that relied on examining nucleotide positions where most people today have a derived allele but where a Neanderthal individual had the ancestral, ape-like allele. In cases where a different DNA fragment from the same or another Neanderthal individual didn't turn out to carry the ancestral allele, Philip suggested that we take a mathematical approach and model the likelihood that this was due to either normal variation among Neanderthals, sequencing errors, or contamination from present-day humans. When Philip later implemented this, the extent of contamination again turned out to be below 1 percent. We finally had estimates of contamination that I trusted and which showed that the quality of our sequences was excellent!

Martin talked about the Illumina data, which we hadn't yet mapped. It made up more than 80 percent of all the fragments sequenced, or almost 1 billion DNA fragments. Much of the discussion centered on the challenges Udo faced in modifying the computer algorithm so that it would map these fragments quickly on the computer cluster in Germany. Although the analysis of the whole genome would have to wait until Udo had mapped all the fragments, we nonetheless discussed how we would do it. The first question was how different the Neanderthal genome was from that of present-day humans. Answering this seemingly simple question was complicated by the errors in the Neanderthal sequences due either to modifications of nucleotides in the ancient DNA or to errors caused by the sequencing technology. Illumina generated up to one error in every hundred nucleotides. To compensate for this, we had sequenced each ancient molecule many times over. But we still estimated that the errors in the Neanderthal DNA sequences added up to about five times as many as in the "gold standard," the human reference genome sequence. Therefore, if we simply counted how many nucleotides differed between the Neanderthal and human genomes, we would be counting mostly errors in our Neanderthal genome.

Ed had a way around this problem. It relied on disregarding all differences that were seen only in the Neanderthal fragments and instead scoring what the Neanderthal carried at positions where the human genome had changed and now differed from the ape genomes. To do this, he simply found all the positions where the human genome differed from the chimpanzee and macaque genomes. He then checked whether the Neanderthal carried the modern human-like nucleotide or the ape-like nucleotide at those positions. If the Neanderthals carried the modern human-like nucleotide, the mutation that caused it was old and predated the split between the Neanderthal DNA fragment and the human reference genome. If the Neanderthals carried the ape-like nucleotide, the mutation was recent and happened in humans after they split from the Neanderthals. Thus, the percentage of substitutions where the Neanderthal was "ape-like" as a fraction of all substitutions along the human lineage gave an estimate of how far back along the human lineage the Neanderthal DNA sequences split from DNA sequences in humans today. The answer was 12.8 percent.

If we assume that our common ancestor with chimpanzees lived 6.5 million years ago, this would mean that the last men and women to transmit their DNA sequences both to people living today and to Neanderthals lived 830,000 years ago. When Ed did the same calculation for pairs of people living today, their common DNA ancestors were found to have lived about 500,000 years ago. So Neanderthals were clearly more distantly related to people living today than people today are to one another; in other words; the Neanderthals are about 65 percent more distantly related to me than I was to another person in the room in Dubrovnik. I could not stop myself from secretly peeking at some of my friends in the sun-lit room and imagine a Neanderthal sitting among us. For the first time I now had a direct genetic estimate of how much closer I was to one of them than to a Neanderthal.

The biggest question on everyone's mind was whether or not Neanderthals and modern humans had interbred. This was David's question to answer, and although he hadn't been able to join us in Dubrovnik, he explained his analyses that suggested interbreeding over a speaker phone. We discussed his results not only in the sessions but throughout coffee breaks and long, lavish and delicious Mediterranean meals that our hosts had organized. The question even dominated the morning runs that Johannes and I took on the outskirts of Dubrovnik, distracting us from the city's medieval beauty and the damage suffered during the recent Balkan war, although not so much that we failed to stick to the paved roads to avoid land mines. Our conversations invariably centered on the intimate relations that may have taken place between modern humans and Neanderthals, who until 30,000 years ago had lived in the very area where we were jogging.

One thing that worried us was that all of our admixture analyses relied on Nick's count of nucleotide matches between the Neanderthal data and either African, European, or Chinese individuals. That left us vulnerable to error in Nick's computer code, the products of which Nick himself was the first to stress that we needed to check. An error could come from some subtle but systematic differences in the techniques used to sequence the modern humans, or from the way Jim Mullikin had mapped them to the human reference genome to find SNPs. The effects of error could be great even if the errors were small; after all, we were talking about differences of only 1 or 2 percent.

During the sessions we collected a list of things to do to check Nick's and David's results. Jim would align his modern human sequences to the chimpanzee genome instead of the human genome to eliminate any bias that might come from the fact that the human reference genome came partly from a European and partly from an African individual. But we also felt that we needed to generate our own DNA sequences from present-day humans. By doing so, we could be certain that they were all produced and analyzed in exactly the same way. Accordingly, if there were systematic problems in our process, we could be certain that the sequences had the same types of errors in them. We decided to sequence the genomes from one person from Europe and one from Papua New Guinea. That might seem an odd choice, but it was prompted by the intriguing observation that we saw an admixture signal that was as strong in China as in Europe. Conventional wisdom had it that Neanderthals had never been in China, but I have always been ready to question paleontological conventional wisdom. Maybe there had been what I liked to call "Marco Polo Neanderthals" in China? After all, Johannes had shown in 2007 that Neanderthals—or at least humans carrying Neanderthal mtDNA—had lived in southern Siberia, some 2,000 kilometers further east of where paleontologists had thought they lived. Maybe some of them had made it to China? However, we were sure that there had never, ever been any Neanderthals in Papua New Guinea, so if we saw the admixture signal there, too, then Neanderthal genes had entered the ancestors of Papuans before they came to Papua, and presumably before Chinese and Europeans separated from each other. We also included a West African, a South African, and a Chinese person in our sequencing plans. With the genomes of these five individuals, we would then do all the analyses again to see if the results held up.

The Dubrovnik meeting ended with a culinary feast that lasted for hours and left us all full of excellent food and pleasantly inebriated. During my career I had been part of many collaborations but none had been as good as this one. Still, I felt a sense of great urgency to bring the project to completion. During dinner, I impressed on everyone that we were now on a tight time schedule, both because the world was awaiting our results after the announcement at the AAAS meeting and because we didn't know what Eddy Rubin was doing in Berkeley with the Neanderthal bones we knew he had collected. Although I hardly ever have bad dreams, I claimed in my improvised speech at the dinner that I had had nightmares about a paper from Berkeley appearing a week before ours with all the same insights we had found.

The next morning I slept on the plane back to Germany. Shortly after returning to Leipzig, I came down with a cold, which developed into a fever and then chest pains in sync with my breathing. I went to the hospital and was diagnosed with pneumonia and given a prescription for antibiotics. But shortly after I got home I received a call to return to the hospital immediately. The lab results suggested I had blood clots somewhere in my system. I soon found myself staring at a CT scan showing blood clots clogging large parts of my lungs. It was a rattling experience. If these clots had reached my lungs as one large clump instead of several smaller pieces, I would have died instantaneously. The doctors blamed the blood clots on too much flying and perhaps the long, cramped bus trip through the night from Split to Dubrovnik. I was prescribed anticoagulants for six months and began researching therapeutic alternatives with the intensity that only comes from being personally affected. To my amazement I stumbled upon references to my father's work from 1943. He had elucidated the chemical structure of heparin, the drug the doctors had given me when I entered the hospital and which had perhaps saved my life. While I found this amusing, I was also quite shaken. It threw a stark light on my family background. I had grown up as the secret extramarital son of Sune Bergström, a well-known biochemist who had shared the Nobel Prize in 1982 for the discovery of prostaglandins, a group of natural compounds that have many important functions in our bodies. I had seen him only occasionally during my adulthood, and the fact that he had worked on the structure of heparin was just one of the innumerable things I did not know about him. The sadness I felt for not having known my father made me realize even more strongly that I wanted to be there when my own three-year-old son grew up. I wanted him to know me. And I wanted to see the Neanderthal project through to completion. It was too early for me to die.

Chapter 18
Gene Flow!

We began sequencing our five modern genomes in May 2009. The pristine DNA, free of the bacterial contamination and chemical damage that marred our Neanderthal samples, yielded about five times as many DNA sequences from each of the five people as we had generated from the Neanderthal. Only a year or two earlier, sequencing those genomes in Leipzig had been unimaginable, but the sequencing technologies such as those marketed by 454 and Illumina had now made it possible for small research groups like ours to sequence several complete human genomes in just a few weeks.

Using the approach he had described in Dubrovnik, Ed estimated how long ago the five present-day human genomes had shared common ancestors with the human reference genome. He found that the European, Papuan, and Chinese individuals shared common ancestors with the reference genome a little over 500,000 years ago. Adding the San from South Africa to the group pushed the point of divergence back to almost 700,000 years ago. The divergence between the San (and related groups) and other people in Africa and elsewhere was among the deepest seen between present-day people. This put the 830,000 year age estimate for the common ancestor of the Neanderthal and present day-human genomes in perspective: having diverged only 130,000 years earlier, Neanderthals were different from us, but not by a lot.

Such calculations must be treated carefully, as they give one single value for the age to a common ancestor as if that was true for the entire genome. Genomes are not inherited as units, however, which means that each part of an individual's genome has its own history and therefore its own common ancestor with the genome of any other individual. This is because each person carries two copies of each chromosome and one of these is independently passed on to a child. So each chromosome has its own independent pattern of history—or its own genealogy, if you will. In addition, each chromosome pair exchanges pieces with each other in an intricate molecular dance called recombination that takes place when egg cells and sperm are formed. Therefore, not only does each chromosome in a population have its own genealogy but each piece of each chromosome does, too. Thus, the ages that Ed had calculated for common ancestors with the reference human genome, 830,000 years for the Neanderthal and 700,000 years for the San, represent grand averages across all parts of the genome.

In fact, when we compared DNA regions from two present-day people with each other, we could easily find regions where they shared a common ancestor just a few tens of thousands of years ago but also regions where they last shared an ancestor 1.5 million years ago. The same was true in a comparison of present-day people with the Neanderthals. So if someone could take a walk down one of my chromosomes and compare it to both a Neanderthal and a reader of this book, that chromosomal pedestrian would find that sometimes I would be more similar to the Neanderthal than to the reader, sometimes the reader would be more similar to the Neanderthal, and sometimes the reader and I would be more similar to each other than to the Neanderthal. Ed's average simply meant that there are slightly more regions of the genome where the reader and I are more similar to each other than either of us are to the Neanderthal.

It is also important to realize that 830,000 years ago is the average age when DNA sequences in people living today have common origins with the DNA sequences carried by the Neanderthal fossils. At that time, these DNA sequences existed in a population whose descendants would eventually give rise to the ancestors of Neanderthals as well as the ancestors of present-day humans. But this was *not* the time when the populations that were to become modern humans and Neanderthals split from each other. That must have happened later. The reason for this is that when we trace the history of DNA sequences in a present-day human and a Neanderthal back in time, the two lineages enter the last population ancestral to both modern humans and Neanderthals—the population where the split between the two groups first happened—and then enter the variation that existed in that ancestral population. So the 830,000 years is a composite age that includes *both* the time when modern humans and Neanderthals were separate populations and the genetic variation that existed in their common ancestral population.

The ancestral population is still totally mysterious to us, although we think it lived in Africa and that some of its descendants eventually left Africa to become the ancestors of Neanderthals. Those who stayed behind were to become the ancestors of people who live today. Estimating when those two groups split using differences in DNA sequences is a tricky proposition, much more so than estimating the time when DNA sequences shared common ancestors. For example, if the population ancestral to Neanderthals and people today contained a lot of variation, more of the DNA sequence differences we found would have accumulated in the ancestral population rather than after Neanderthals and modern humans went their separate ways. This would make the population split relatively recent. We were able to crudely estimate the level of variation in the ancestral population from how different the time estimates to common DNA ancestors were for different segments of the genome. To estimate the population split time, we also needed to know the generation time, or the average age at which individuals produced offspring, something that we obviously didn't know. Taking these uncertainties into account as best we could, we came to the conclusion that the population split seemed to have happened sometime between 270,000 and 440,000 years ago, although even that might also underestimate the uncertainty. Nevertheless, the ancestors of people today and the ancestors of Neanderthals probably went their separate ways at least 300,000 years ago.

Having gauged how different Neanderthals and modern humans were, we returned to the question of what happened when the ancestors of present-day people left Africa and met their long-lost Neanderthal "cousins" in Europe. To see whether those modern humans and Neanderthals exchanged genes, Ed quickly mapped our five human genomes to the chimpanzee genome and David and Nick repeated their analyses. I was convinced that the results would now be reliable, and I secretly suspected that the extra similarity between the Neanderthals and the Europeans and Chinese would disappear.

On July 28, I received two long e-mails from David and Nick. It is a testimony to David's passion for science that the analyses went forward even though his wife Eugenie gave birth to their first child on July 14. Nick had done the ten possible pairwise comparisons among the five modern human genomes. In each case, he identified SNPs where a chromosome in one individual differed from a chromosome in the other. He found about 200,000 such differences between any one pair, more than enough SNPs to accurately determine whether the Neanderthal was closer to one human or the other.

Nick found that the Neanderthals matched the San in 49.9 percent and the Yoruba in 50.1 percent of cases. This was expected since Neanderthals had never been in Africa and therefore should not have more of a relationship to some Africans than to others. When he used SNPs where the French differed from the San, the Neanderthals matched the French in 52.4 percent of cases. We now had such a vast amount of data that there was only a 0.4 percent uncertainty in these values. So it was very clear that the French genome was more similar to the Neanderthals than was the San. For the comparison using the French and the Yoruba, the corresponding value was 52.5 percent. For SNPs where the Chinese differed from the San and the Yoruba, the values were 52.6 percent and 52.7 percent, respectively, and for SNPs where the Papuans differed from the Africans, it was 51.9 percent and 52.1 percent. When he analyzed SNPs where the French, Chinese, and Papuans differed among each other, the values varied between 49.8 percent and 50.6 percent. So in all comparisons between people that did not involve Africans, the values were around 50 percent. But whenever an African and a non-African were compared, the Neanderthal matched the non-African at around 2 percent more SNPs than did the African. There did indeed seem to be a small but clearly discernible genetic contribution from Neanderthals to people outside Africa, no matter where they lived.

I read the two e-mails once. Then I read them again, this time very carefully, trying to catch any hint of a flaw in the analyses. I could find none. I leaned back in my office chair and looked blankly at my very untidy desk where papers and notes from the past few years had accumulated in layer after layer. David and Nick's results stared at me from the computer screen. This was not a technical error of some sort. Neanderthals *had* contributed DNA to people living today. It was amazingly cool. It was what I had dreamed of achieving for the last twenty-five years. We had hard evidence to answer a fundamental question debated for decades about human origins, and the answer was unexpected. By showing that not all of the genomic information in present-day humans traced back to recent ancestors in Africa, it contradicted the strict out-of-Africa hypothesis of which my mentor Allan Wilson had been one of the main architects. It contradicted what I myself had believed to be true. Neanderthals weren't totally extinct. Their DNA lived on in people today.

Staring blankly at my desk I realized that our results were unexpected not only in that they contradicted the out-of-Africa hypothesis. They also didn't support the common version of the multiregional hypothesis. Contrary to the predictions of this hypothesis, we didn't see the Neanderthal genetic contribution only in Europe, where Neanderthals had lived. We saw it also in China and Papua New Guinea. How could this be? Absent-mindedly, I started to clean my desk. Slowly at first but then with increasing energy, I tossed out debris from years-old projects. Dust whirled into the air from layers deep on my desk. I needed to start a new chapter. I needed a clean desk.

Doing domestic tasks sometimes helps me think and as I cleaned, I visualized modern humans as arrows on a map coming out of Africa and meeting Neanderthals in Europe. I could imagine them having babies with Neanderthals—babies who then became incorporated among the modern humans, but I struggled to see how their DNA came to East Asia. It was possible that subsequent migration among modern humans might have brought Neanderthal DNA to China, but it seemed we would then find less similarity on average between a Chinese person and Neanderthals than between a European person and Neanderthals. Then it dawned on me: my imaginary arrows showing modern humans coming out of Africa passed

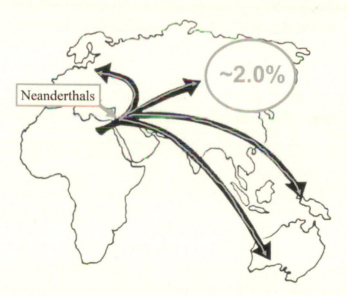

FIGURE 18.1. An illustration of the idea that if the Neanderthals mixed with early modern humans leaving Africa, and these went on to populate the rest of the world outside Africa, they would carry Neanderthal DNA with them to regions where Neanderthals never existed. For example, about 2 percent of the DNA of people even in China comes from Neanderthals. Photo: Pääbo, MPI-EVA.

through the Middle East! This was of course the first place where modern human would have met Neanderthals. If those humans mixed with the Neanderthals and then went on to become the ancestors of all people outside Africa today, the result would be that everyone outside Africa would carry approximately the same amount of Neanderthal DNA (see Figure 18.1). This must be a possible scenario. But I knew from experience that my intuition could sometimes be very wrong. Fortunately, I also knew that people like Nick, David, and Monty, who tested ideas mathematically, would set me straight if it was.

We discussed David and Nick's findings at our Friday meetings and during intense weekly consortium phone meetings. While some of us were now convinced that Neanderthals had mixed with modern humans, others were still reluctant to believe it, even though they struggled to explain how David and Nick's analysis could possibly be wrong. I realized that if it was that hard to get everyone in our consortium to believe these results, we would have an even harder time convincing the world, especially the many paleontologists who saw no evidence for interbreeding with Neanderthals in the fossil record. These included some of the most respected people in the field, like Chris Stringer at the Natural History Museum in London and Richard Klein at Stanford University in California. Although I thought these paleontologists stood for a properly cautious interpretation of the fossil record, it still seemed possible that they had been influenced by previous genetic results. Many groups, including ours, had shown that the big picture of genetic variation in people today was one in which genetic variation had come out of Africa rather recently. Our 1997 paper showing that Neanderthals had not contributed any mtDNA to people today had also had a big influence. Although some paleontologists, such as Milford Wolpoff at the University of Michigan and Erik Trinkaus at Washington University in St. Louis, saw evidence of mixing in fossils, and some geneticists had made attempts to point out gene variants that might have come from Neanderthals, such arguments weren't compelling enough to sway the common opinion. Or at least they hadn't struck me as compelling. There had simply never before been a need to invoke any Neanderthal contribution to explain the patterns of either morphological or genetic variation across the world today. Now this situation had changed. We could look directly at the Neanderthal genome. And we saw a contribution, albeit a small one.

Still, I suspected we would need more to convince the world of our results. Science is far from the objective and impartial search for incontrovertible truths that nonscientists might imagine. It is, in fact, a social endeavor where dominating personalities and disciples of often defunct yet influential scholars determine what is "common knowledge." One way to undermine this aspect of common knowledge would be to do additional analyses of the Neanderthal genome, independent of the counting of SNP alleles that David and Nick had done. If such additional and independent lines of evidence also suggested gene flow from Neanderthals into modern humans, then the world at large would be easier to convince. Finding what other analyses we could do became a constant theme in our weekly phone meetings.

Somewhat unexpectedly, a viable suggestion came from outside our consortium. At the Cold Spring Harbor meeting in May 2009, David had met with Rasmus Nielsen, a Danish population geneticist who had done his PhD with Monty Slatkin back in 1998. He was now a professor of population genetics at UC Berkeley. Rasmus told David that he and his postdoc Weiwei Zhai had searched present-day genomes for regions that showed greater variation *outside* Africa than *inside* Africa. Although certainly possible, such a pattern is generally unexpected, as initially small offshoots of larger populations generally contain just a subset of the variation found in the ancestral group. If any such regions were found, there could be many explanations, but one possibility interested us very much. Since Neanderthals had lived independently from modern human ancestors for a few hundred thousand years outside Africa, they must have accumulated genetic variants distinct from those in modern humans. If they had subsequently contributed segments of their genome to people outside Africa, then Rasmus's approach might identify genomic regions where this had happened because in such regions there should be just that pattern of more variation outside Africa than inside Africa. Using our Neanderthal genome, we could now check whether at least some of these regions came from Neanderthals, since the non-African versions of Rasmus's regions would then be close to our Neanderthal DNA sequence. In June 2009, I asked Rasmus and Weiwei to join the Neanderthal Genome Analysis Consortium.

Rasmus focused on the regions that were particularly unusual in having deep divergences in Europe relative to Africa. There were seventeen such regions. Ed sent Rasmus the Neanderthal DNA sequences we had in hand from fifteen of these seventeen regions. He wrote back in July with stunning results. At thirteen of the fifteen regions, the Neanderthals did indeed carry the variants now found in Europe and not in Africa. Later, Rasmus refined his analysis and focused on twelve regions that were over 100,000 nucleotides long and found that the Neanderthals carried the variants now found in Europe for ten of these regions. This was indeed an amazing result! I could not imagine any other explanation for this than gene flow from Neanderthals into people outside Africa. Although this was what scientists call a qualitative result, as it didn't allow us to calculate how much DNA Neanderthals had contributed to people in Europe or Asia, it made vividly clear that such a contribution had occurred. And it was an independent line of support of David and Nick's other, quantitative, analyses that had led to the same conclusion.

We continued to think about additional ways to test for gene flow. As always, David tended to come up with brilliant ideas. He argued that a trivial reason why a region of the genome today might be close to a Neanderthal is that it accepts very few mutations, either because its mutation rate is low or because it cannot mutate without causing a person to die. If a genomic region in me is similar to a Neanderthal for this reason, then one would expect that region to also be close to other humans today, simply because it rarely changes. But if a region in me is similar to a Neanderthal because my ancestors had inherited it from Neanderthals, then there is no reason why I should be close to other people. In fact, I may even tend to be far from others thanks to the distinct evolutionary history of Neanderthals.

David set out to implement these insights into our analysis. He used the European parts of the human reference genome and divided them into segments. He then plotted the number of differences in those segments from the Neanderthal genome and from another European genome (that of Craig Venter). In general, he saw that the closer the European segments of the reference genome were to the Neanderthals, the closer they were to Craig's genome, suggesting that the rate with which segments accumulated mutations determined the differences both to the Neanderthal genome and to Craig's genome. But when he came to segments where the European was *very* similar to the Neanderthals, the relationship reversed and suddenly became *more* different from Craig's genome. I was already convinced from the other analyses that gene flow had occurred. But when David presented these results during a visit to our lab in December 2009, I felt sure that we would be able to convince the world that Neanderthal DNA segments were lingering on in people today. No matter how we looked at the data, we came up with the same result.

We could now turn our full attention to how, when, and where modern humans had interacted intimately with Neanderthals. The first question was the direction in which the gene flow had occurred—that is, whether modern humans had contributed DNA to Neanderthals, Neanderthals had contributed DNA to modern humans, or both. Although one might think that genes would flow equally in both directions when two human groups meet, in real life that is rarely the case. Often one group is socially dominant over the other. A common pattern is then that men from the dominant group sire children with women from the nondominant group and the children remain with their mothers in the nondominant group. Thus, gene flow will tend to be from the socially dominant to the nondominant group. Obvious examples are white slave owners in the American South and British colonialists in Africa and India.

We tend to think that modern humans were dominant over Neanderthals, as Neanderthals eventually disappeared. But our data actually suggested that gene flow had been *from* Neanderthals *into* modern humans. David's last result, for example, showed that the DNA regions where some Europeans were similar to Neanderthals tended to be very different from those found in other Europeans. The implication was that these regions had accumulated differences separately from those of other Europeans for some time before entering the present European gene pool. Presumably this had happened in the Neanderthals. If the contribution had gone in the other direction—from modern humans into Neanderthals—those regions would just have been average parts of the genome with average amounts of differences from other Europeans. For this and other reasons, we concluded that all, or almost all, of the gene flow was from Neanderthals into modern humans.

This did not necessarily mean that children from Neanderthal-modern human unions were never raised by Neanderthals. In 2008, Laurent Excoffier, a Swiss population geneticist who had always taken an interest in data generated by our group, published a paper about gene flow between two populations, after which one expands while the other one doesn't, or even shrinks in size. In such a case the gene variants exchanged between the populations are more likely to be preserved in the growing population than in the dwindling population. And if the contribution occurs along the "wave front" of an advancing population, where the advancing population is in the process of expanding, then the contributed variants might even reach quite high frequencies. Excoffier had aptly named this phenomenon "allelic surfing," illustrating the fact that an allele that entered the advancing "wave" of a colonizing population might surge to high frequencies. This meant that interbreeding might have happened in both directions but that we wouldn't detect it in Neanderthals, because, after the encounter, their population size was likely to have shrunk.

Another more mundane reason why we might not detect gene flow from modern humans into Neanderthals is that the 38,000-year-old Neanderthals from Vindija Cave simply lived before interchange happened. Perhaps we will never really know the details of how Neanderthals and modern humans had interbred, but I am not overly troubled by this. To me, "who had sex with whom" in the Late Pleistocene is a question of secondary importance. What matters is that Neanderthals did in fact contribute genes to people today. That is what matters with respect to the genetic origin of people today.

Having confirmed David and Nick's findings, we explored the question of how much of the genomes of people outside Africa had come from Neanderthals. This could not be directly estimated from the SNP matching frequencies because the number of extra matches between Neanderthals and people outside Africa depended on a number of other variables. One variable was when the common ancestor of Neanderthals and modern humans lived, another was when they had mixed with each other, and a third was how large the populations of Neanderthals had been. Monty Slatkin estimated the fraction of Neanderthal DNA in present-day people by modeling the population history of Neanderthals and modern humans. His results suggested that people who are of European or Asian ancestry have inherited between 1 and 4 percent of their DNA from Neanderthals. David and Nick did a different analysis where they essentially asked how far Europeans and Asians are toward being 100 percent Neanderthals. The answer varied between 1.3 and 2.7 percent. Thus we concluded that less than 5 percent of the DNA of people outside Africa came from Neanderthals—a small but clearly discernible proportion.

The final question for that round of work was how Neanderthal DNA wound up not only in Europeans but also in the Chinese and the Papuans. As far as we knew, Neanderthals had never been in China and surely they had never made it to Papua New Guinea, which led us to infer that

Neanderthals and the ancestors of the Chinese and the Papuans must have met somewhere farther west.

I kept my Middle Eastern idea to myself as we sat huddled around the speaker phone in my Leipzig office during our weekly phone meeting so that the sharp minds of the people in the consortium would explore all possibilities. Monty came up with a complicated scenario to explain the patterns in variation we saw. First, he assumed that Neanderthal ancestors originated in some corner of Africa and then left Africa and evolved into Neanderthals in western Eurasia sometime around 300,000 to 400,000 years ago. Second, if the place where Neanderthal ancestors originated in Africa was the same as the place where 200,000 or more years later the ancestors of modern humans originated; and if populations in Africa over that time had remained subdivided so that there were differences in allele frequencies that persisted from the time the Neanderthal ancestors left until the time the modern human ancestors started spreading; and if, when modern humans originated, they swept not only out of Africa but also across Africa and incorporated variants that existed there through interbreeding with archaic African humans, then the result would be that Neanderthals were more similar to people outside Africa than inside Africa, just as we were seeing.

Although this scenario was theoretically possible, it required the persistence of a stable subdivision of the Africa population for hundreds of thousands of years. As Monty himself pointed out, this seemed unlikely since humans are quite prone to moving around. The bigger problem was its complexity. For reconstructing the past, it is considered best to favor the simplest scenario that can account for the patterns seen, even though many other, more complex scenarios are also possible. The principle of preferring the simplest explanation is called parsimony. For example, others had assumed that the ancestors of modern humans and Neanderthals had originated in Asia and that the ancestors of modern humans went to Africa, leaving no descendants in Eurasia, and then expanded out again to replace the Neanderthals. This hypothesis is indeed compatible with all observations, but it requires more population movements and population extinctions than the simpler assumption that Neanderthals had originated in Africa. So the Asian-origin scenario is less parsimonious, and therefore inferior as an explanation, than the African-origin scenario. Thus we noted the African substructure scenario as a possible explanation for our data but regarded it as unlikely, since there was a simpler, more obvious explanation that was in fact so obvious that several of us had thought of it independently: the Middle Eastern scenario.

Chapter 19
The Replacement Crowd

The earliest remains of modern humans yet found outside Africa have been discovered in the Carmel mountain range in Israel. Bones found in two caves named Skhul and Qafzeh are more than 100,000 years old. And at two other sites just a few kilometers away, Tabun Cave and Kebara Cave, skeletons of Neanderthals that are about 45,000 years old have been found. These finds don't necessarily mean that Neanderthals and modern humans lived side by side in the Carmel Mountains for over 50,000 years; in fact, many paleontologists suggest that modern humans from the south lived in the area when the climate was warmer and that Neanderthals from the north moved in, and the modern humans moved out, during colder periods. It has also been suggested that the modern humans from Skhul and Qafzeh died out without leaving any descendants. But even if they did not leave descendants, they probably had relatives. And even if they weren't constant neighbors, the two groups must have made contact, over periods of thousands of years, even as changes in climate might have shifted the zone of contact, sometimes north, sometimes south. That, in brief, is the Middle East scenario.

The Middle East, as I learned from talking to paleontologists, particularly Jean-Jacques Hublin, the French scientist who joined our institute as the director of the department of human evolution in 2004, was an attractive place for modern humans and Neanderthals to have mixed 50,000 to 100,000 years ago. One reason is that it is the only area in the world where we know that Neanderthals and modern humans were, at least potentially, in contact for such a long time. Another reason is that neither of the two groups appears to have been clearly dominant during that period. For example, the stone tools used by both groups were the same. In fact, since their tool kits are identical, the only way one can know for certain whether a Middle Eastern archaeological site from that time period was occupied by Neanderthals or modern humans is if skeletal remains are present.

This all changed shortly after 50,000 years ago. At that time, modern humans firmly established themselves outside of Africa and began to rapidly fan out over the Old World, reaching Australia in just a few thousand years. By then, the way in which they interacted with Neanderthals seems to have changed. In Europe, where the fossil record is particularly well studied, it seems clear that when modern humans appeared in an area, Neanderthals disappeared either immediately or shortly thereafter. The same eventually happened all around the world: wherever modern humans appeared, earlier forms of humans, sooner or later, disappeared.

To distinguish these expansive and ambitious modern humans from the modern humans who were hanging around in Africa and the Middle East between 100,000 and 50,000 years ago, I like to call them the "replacement crowd." They had developed a more sophisticated tool culture, called the Aurignacian by archaeologists, characterized by flint tools of different sorts, including a variety of blades. Points for spears and arrows made of bone are often found at Aurignacian sites, representing what some archaeologists believe to be the first instances of projectile weapons. If true, this invention, which for the first time allowed humans to kill animals and enemies at a distance, may in itself have tipped the balance in their favor when they met Neanderthals and other earlier forms of humans. The Aurignacian culture also produced the first cave art and the first figurines of animals, including mythical figures that are half-human and half-animal, suggesting that they possessed a rich inner life that they wanted to communicate with others in their group. The "replacement crowd" thus exhibited behaviors that were only occasionally or not at all seen among Neanderthals and among the earlier modern humans from Skhul and Qafzeh.

We don't know where the "replacement crowd" came from. In fact, they could even have been the descendants of the same humans who had already been living in the Middle East, simply accumulating the cultural inventions and proclivities that enabled "replacement," but it is more likely that they came from somewhere in Africa. In any event, the "replacement crowd" must have spent time in the Middle East.

As the replacement crowd moved into the Middle East they may well have incorporated the modern humans already there into their groups. Those humans, in turn, may have already mated with Neanderthals, so that the Neanderthal DNA passed through them into the replacement crowd, and then on to us today. Such a model may seem more complex, and thus less parsimonious, than is ideal. A major problem facing a direct model, where the replacement crowd mated with Neanderthals, is the question of why, if they were willing to rear children with Neanderthals in the Middle East, the replacement crowd didn't also do so later when they met and replaced them in central and western Europe. If they had, Europeans should have more Neanderthal DNA than Asians. What this indirect scenario suggests is that perhaps the replacement crowd never mixed with Neanderthals but instead received the Neanderthal contribution through these other modern humans, those whose remains have been found in Skhul and Qafzeh. These very early modern humans, having a culture very similar to Neanderthals and having lived next to them for tens of thousands of years, may have been more inclined to mate with Neanderthals than to "replace" them.

This indirect model is admittedly pure speculation. It may be that we don't see an additional contribution in Europe because it is too small for us to detect. It may also be that the interbreeding in the Middle East was followed by a particularly large growth of the population that had mixed with the Neanderthals. If so, this would mean that we are particularly able to detect that event because of the "surfing" that Excoffier had described, and less able to detect later events that were not followed by any equally large population expansion. Or perhaps there was later migration from Africa into Europe that "diluted" the extra Neanderthal contribution in Europe. I hope that direct evidence in the future can address this. Should it prove possible to study the DNA from the Skhul and Qafzeh people, then it will be possible to see if they had mixed with the Neanderthals, perhaps to a large extent. It should then also become possible to see if they carried the very same Neanderthal DNA fragments that today exist in people in Europe and Asia.

For the moment, the simplest—the most parsimonious—scenario is that the replacement crowd met and mated with Neanderthals somewhere in the Middle East and raised the children who resulted from those unions. Those children, part modern human, part Neanderthal, became members of the replacement crowd, carrying the DNA of the Neanderthals and their descendants as an internal fossil of sorts. Today, those internal Neanderthal remains have reached even the southern tip of South America, Tierra del Fuego, and Easter Island in the middle of the Pacific. The Neanderthals live on in many of us today.

When we had come this far in our analyses, I began to worry about what the social implications of our findings might be. Of course, scientists need to communicate the truth to the public, but I feel that they should do so in ways that minimize the chance for it to be misused. This is especially the case when it comes to human history and human genetic variation, when we need to ask ourselves: Do our findings feed into prejudices that exist in society? Can our findings be misrepresented to serve racists' purposes? Can they be deliberately or unintentionally misused in some other way?

I could imagine a few scenarios. It is generally not a compliment to be called a "Neanderthal," and I wondered if some individuals might link Neanderthal DNA with aggressiveness or other behaviors associated with the colonialist expansion out of Europe. I didn't see this as too much of a threat, however, as such "reverse racism" against Europeans would most likely not be very virulent. A more serious issue was what it meant that Africans lacked this component. Were they not part of the "replacement crowd"? Was their history somehow fundamentally different?

Reflecting on such questions, I came to realize that this was unlikely to be the case. The most reasonable scenario is that all humans today, regardless of whether we live inside or outside of Africa, are part of the replacement crowd. And although many paleontologists and geneticists, including myself, had thought that the replacement crowd spread around the world without mixing with the other human groups that they encountered, after its having happened once was ascertained, there was reason to think it may have happened more times. Since we have no ancient genomes from other parts of the world, we are effectively blind to possible contributions from other archaic humans. This is particularly the case in Africa, where genetic variation is larger than elsewhere, so a contribution from some archaic group would be hard to detect. Nevertheless, when the replacement crowd spread across Africa, they could well have mixed with archaic humans there and incorporated their DNA into their gene pool. I decided to point this out to journalists and in talks to make it clear that there was little reason to believe that Africans had no archaic DNA in their genomes. Probably all humans do, and indeed, some more recent analyses of present-day people in Africa have suggested that this is the case.

One evening, when I was especially tired after a long day at work followed by some particularly rambunctious behavior by our now five-year-old son, a crazy question came to me just after he had fallen asleep: If all people today carry 1 to 4 percent of the Neanderthal genome, could one imagine that by pure chance, as a freak result of the random assortment of DNA segments during the production and fusion of sperm and eggs, a child could be born who is entirely or almost entirely Neanderthal? Could the many Neanderthal DNA fragments that exist in people today have happened to come together in my sperm cell and Linda's egg cell that ended up developing into our rambunctious son? Just how Neanderthal could he—or I—be?

I decided to do a simple calculation. The segments that Rasmus had identified were about 100,000 nucleotides long, and on average perhaps 5 percent of people outside Africa carried any one of them. If all Neanderthal fragments were of this length and if, together, they made up the entire Neanderthal genome, there would be about 30,000 fragments in existence. Many Neanderthal DNA fragments were in fact both shorter in length and less frequent than 5 percent, and perhaps they wouldn't add up to the whole genome, but I wanted to deliberately bias my calculations to see if it could at all be possible that my son was of completely Neanderthal descent. Under these assumptions, his chance of having a particular Neanderthal DNA fragment was like drawing a ticket in a lottery where 5 percent of tickets were winners. His chance of carrying the Neanderthal fragment on both of a pair of his chromosomes was like drawing a winning ticket in this lottery twice. This was 5 percent of 5 percent or 0.25 percent. To be entirely Neanderthal for the genome he had gotten from Linda and the genome he had gotten from me, he would have to have drawn winning tickets twice for each of the 30,000 segments, or 60,000 times in row! The chance of this was of course infinitesimally small (in fact, a zero and a decimal point followed by 76,000 zeros and then some number). So not only was my son very unlikely to be wholly Neanderthal, even among all 8 billion people on earth there was no chance that a Neanderthal child would be born. So I had to dismiss the idea that my son was to any appreciable degree Neanderthal. Thankfully, I could also write off the risk that any latter-day Neanderthal would walk into our lab one day and offer me a blood sample, making our entire effort to sequence a Neanderthal genome from ancient bones unnecessary.

Nevertheless, clearly identifying which DNA segments in our genomes come from Neanderthals and finding out if all parts of the Neanderthal genome exist scattered among people today are both important research goals. The size and number of these segments would say something about how many cases of actual mixed children were behind the contribution of Neanderthal DNA to the replacement crowd and when this contribution occurred. Also, any parts that might be missing could be very interesting, because they may contain the genetic essence of the crucial difference between the modern human replacement crowd and Neanderthals.

At this point in my musings, I realized after making the calculations about my son that others would also be interested in finding out what parts of their genomes were of Neanderthal origin. People wrote to me every year suggesting that they (or their loved ones) were part Neanderthal. Often they included photos, which tended to show slightly stocky individuals, and quite often they volunteered to contribute a blood sample for our research. Now that we actually had a Neanderthal genome, I could imagine comparing our Neanderthal DNA sequences to DNA sequences of any person today and identifying the segments in the person's genome that were close enough to the Neanderthals to have been inherited from them. After all, there were already many companies that offered this kind of analysis with respect to ancestry from different parts of the world. For example, people in the United States are often interested in finding out how much of their ancestry comes from Africa, Europe, Asia, or from Native Americans. In the future, this could be done for Neanderthal ancestry. I was intrigued, but again, I was also worried. There might be a stigma associated with being "Neanderthal." Would people feel bad if they knew that some part of their genome that carried genes involved in how brain cells work came from Neanderthals? Would future arguments between spouses include arguments such as "You never take out the trash because such-and-such brain gene of yours is Neanderthal"? Could this stigma be applied to entire groups of people if some population happened to have a high frequency of a Neanderthal variant of a gene?

I felt that we should try to control such applications of our work. The only way to do this that I could think of was to patent the use of the Neanderthal genome for such ancestry testing. If we did so, anyone who wanted to earn money by testing people would need to obtain a license from us. That would allow us to impose conditions on how information was given to the customers. We could also charge a fee for such licenses so that our lab and the Max Planck Society might get back some of the money invested in the Neanderthal project. I talked about this to Christian Kilger, a former graduate student who was now an attorney specializing in biotech patents in Berlin. Together we discussed how one could share putative patent revenues among the research groups in the consortium.

Thinking this plan might be slightly controversial, I presented it to the group in one of our Friday meetings. I soon found out that I had totally misjudged the situation. Some people were passionately against the idea of a patent. In particular, Martin Kircher and Udo Stenzel, whose professional abilities I much respected, were against patenting the use of something that occurred naturally, such as the Neanderthal genome. Overall, this was a minority view in the group but it was upheld with almost religious fervor. Others held the exact opposite viewpoint. Ed Green, for example, had even visited the largest commercial ancestry company, 23andMe, in California and seemed open to working with it in the future. The debate raged in our meetings, in the cafeteria, in the labs, and at our desks. I invited Christian Kilger and a patent attorney from the Max Planck Society to explain what patents were and how they functioned. They went to great lengths to explain that a patent would put limitations only on the commercial use of the Neanderthal genome—and even then, only for the particular purpose of ancestry testing—and that it would in no way limit any scientific applications. This did nothing to change any opinions or end the emotional tone of our debate.

I didn't want a long divisive fight about this issue in the group. I wanted even less to push through a decision against the will of a dedicated minority. We were still far from submitting our paper and needed cohesion in the group. So, two weeks after raising the issue, I announced during a Friday meeting that I had decided to drop the patent idea. I received an e-mail from Christian that ended with "What a chance missed." I shared his sentiment. It had been an opportunity to both fund future research and positively influence how commercial companies could use our results. In fact, as I write this, 23andMe has started offering Neanderthal ancestry testing. Other companies are sure to follow. But group cohesion was what drove our project forward. It was too valuable an asset to risk destroying.

Chapter 20

Human Essence?

Our institute in Leipzig is a fascinating place. In one way or another, almost every researcher there studies what it means to be human, but they all approach this rather fuzzy-sounding question from a fact-oriented, experimental perspective. One particularly interesting line of research is that of Mike Tomasello, the director of the department for comparative and developmental psychology. His group is interested in differences in cognitive development between humans and the great apes.

To measure those differences, Mike's group administers the same "intelligence" tests to both. Of special interest is how well apes and human children cooperate with their peers to achieve goals such as figuring out how to get an intricate contraption to release a toy or candy. One insight that has come from Mike's work is that, until about ten months of age, there are hardly any detectable cognitive differences between young humans and young apes. However, at around one year of age, humans start doing something that the ape youngsters don't: they start to draw others' attention to objects of interest by pointing at them. What's more, from that age on, most human children find pointing at things intrinsically interesting. They will point to a lamp, a flower, or a cat, not because they want the lamp, the flower, or the cat, but for the sole purpose of directing the attention of their moms, dads, or others to it. It is the very act of directing the attention of another person that is fascinating to them. It seems that by about one year of age, they have begun both to discover that other people have a worldview and interests not so dissimilar from their own and to take steps toward being able to direct the attention of others.

Mike has suggested that this compulsion to direct the attention of others is one of the first cognitive traits that emerge during childhood development that is truly unique to humans.[1] It is certainly one of the first signs that the children have started to develop what psychologists call a theory of mind, an appreciation that others have different perceptions than one's own. It is easy to imagine that the enormous human capacity for social activities, for manipulating others, for politics, and for concerted action of the sort that result in large and complex societies arise out of this ability to put oneself in another's shoes and manipulate that person's attention and interest. I believe that Mike and his group have pointed to something that is fundamental for what set humans on a historic trajectory so different from that of the apes and the many extinct forms of humans, such as Neanderthals.

Mike has also pointed out another potentially very important propensity that sets human children apart from ape youngsters: human children, much more than apes, tend to imitate what their parents and other humans do. In other words, human children "ape" whereas apes do not "ape." And reciprocally, human parents and other adults correct and modify behaviors in their children to a much greater extent than ape parents do. In many societies, humans have even formalized this activity—it is what we know as teaching. In fact, a very large part of all activities that humans do with their children is teaching, in either an implicit or explicit form. Often it is institutionalized in the form of school and universities. In contrast, there has been almost no teaching observed in apes. It is fascinating to me that the human propensity to readily learn from others may emanate from the shared attention that first manifests itself in the toddler who points to the lamp just to get her dad to look at it.

This focus on teaching and learning probably has fundamental consequences for human societies. Whereas apes must learn every skill they eventually acquire through trial-and-error and without a parent or other group member actively teaching them, humans can much more effectively build on the accumulated knowledge of previous generations. As a result, when an engineer improves a car, she need not invent it from scratch. She will build on the inventions of previous generations all the way back to the invention of the combustion engine in the twentieth century and of the wheel in antiquity. To this accumulated wisdom of her ancestors she will merely add some modifications to the design that later generations of engineers will in turn take for granted and continue to build upon. Mike calls this the "ratchet effect." It is clearly a key to the enormous cultural and technological success of humans.

My fascination with Mike's work stems from my conviction that there are genetic underpinnings to our propensity for shared attention and the ability to learn complex things from others. In fact, there is ample evidence to suggest that genetic traits are a necessary foundation to these human behaviors. In the past, people sometimes did what we now consider to be unethical experiments in which they raised newborn apes together with their own children in their home. Although apes learned how to do many human-like things—they could construct simple two-word sentences, manipulate household appliances, use bicycles, and smoke cigarettes—they did not learn truly complex skills and they did not engage in communication on the scale that humans do. In essence, they did not become cognitively human. So it's clear that there is a biological substrate necessary for fully acquiring human culture.

This is not to say that genes alone are sufficient for acquiring human culture, only that they are a necessary substrate. In the imaginary experiment where a human child is raised in the absence of any contact with other human beings, it is very likely that the child would never develop most of the cognitive traits that we associate with humans, including awareness of the interests of others. That unfortunate child would probably also not develop the most sophisticated of cultural traits that emanates from our tendency to share attention with others: language. So I am convinced that social input is necessary for the development of human cognition. However, no matter how early in life and how intensively they are integrated into human society and no matter how much teaching they are subject to, apes do not develop more than rudimentary cultural skills. Social training alone is not enough. A genetic readiness to acquire human culture is necessary. Similarly, I am convinced that a newborn human raised by chimpanzees would fail to become cognitively chimpanzee. There is surely also a genetic substrate necessary to becoming fully chimpanzee that humans lack. But since we are humans, we are more interested in what makes humans human than in what makes chimpanzees chimpanzee. We should not be ashamed of being "humancentric" in our interests. In fact, there is an objective reason to be so parochial. The reason is that humans, and not chimpanzees, have come to dominate much of the planet and the biosphere. We have done so because of the power of our culture and technology; these have allowed us to increase our numbers vastly, to colonize areas of the planet that otherwise would not have been habitable for us, and to have an impact on and even threaten aspects of the biosphere. Understanding what caused this unique development is one of the most fascinating, perhaps even one of the most pressing, problems that scientists face today. One key to the genetic underpinnings of this development may well be found through comparing the genomes of present-day humans with Neanderthals. Indeed, it is this feeling that kept me going during years of struggling with the technical minutiae of retrieving the Neanderthal genome.

According to the fossil record, Neanderthals appeared between 300,000 and 400,000 years ago and existed until about 30,000 years ago. Throughout their entire existence their technology did not change much. They continued to produce much the same technology throughout their history, a history that was three or four times longer than what modern humans have experienced. Only at the very end of their history, when they may have had contact with modern humans, does their technology change in some regions. Over the millennia, they expanded and retracted with the changing climates in the areas they lived in Europe and western Asia, but they didn't expand across open water to other uninhabited parts of the world. They spread pretty much as other large mammals had done before them. In that, they were similar to other extinct forms of humans that had existed in Africa for the past 6 million years and in Asia and Europe for about 2 million years.

All of this changed abruptly when fully modern humans appeared in Africa and spread around the world in the form of the replacement crowd. In the 50,000 years that followed—a time four to eight times shorter than the entire length of time the Neanderthals existed—the replacement crowd not only settled on almost every habitable speck of land on the planet, they developed technology that allowed them to go to the moon and beyond. If there is a genetic underpinning to this cultural and technological explosion, as I'm sure there is, then scientists should eventually be able to understand this by comparing the genomes of Neanderthals to the genomes of people living today.

Fueled by this dream, I was itching to start looking for crucial differences between Neanderthals and present-day humans once Udo had finally mapped all the Neanderthal fragments in the summer of 2009. But I also realized that I needed to be realistic about what those differences would tell us. The dirty little secret of genomics is that we still know next to nothing about how a genome translates into the particularities of a living and breathing individual. If I sequenced my own genome and showed it to a geneticist, she would be able to say approximately where on the planet I or my ancestors came from by matching variants in my genome with the geographic patterns of variants across the globe. She would not, however, be able to tell whether I was smart or dumb, tall or short, or almost anything else that matters with respect to how I function as a human being. Indeed, despite the fact that most efforts to understand the genome have sprung from efforts to combat disease, for the vast majority of diseases, such as Alzheimer's, cancer, diabetes, or heart disease, our current understanding allows us only to assign vague probabilities to the likelihood that an individual will develop them. So in my realistic moments, I realized that we would not be able to directly identify the genetic underpinnings of the differences between Neanderthals and modern humans. There would be no smoking gun to be found.

Still, the Neanderthal genome was a tool that would allow us to begin to ask questions about what set Neanderthals and humans apart—a tool that not only we but all future generations of biologists and anthropologists would be able to use. The first step was obviously to make a catalog of all the genetic changes that happened in the ancestors of people living today after they separated from the ancestors of the Neanderthals. These changes would be many, and most of them would be without great consequences, but the crucial genetic events that we were interested in would be hidden among them.

The crucial task of making the first version of such a catalog of all changes unique to modern humans was taken on by Martin Kircher together with his supervisor Janet Kelso. Ideally, such a catalog should contain all genetic changes that are present today in all or nearly all humans and that occurred after modern humans parted ways with the ancestors of Neanderthals. The catalog should thus list positions in the genome where the Neanderthal looked like the chimpanzee and other apes while all humans, no matter where they lived on the planet, differed from the Neanderthals and the apes. However, in 2009 there were many limitations to how complete and correct such a catalog could be. First of all, we had sequenced only about 60 percent of the Neanderthal genome so the catalog could only be 60 percent complete. Second, even if we saw a difference from the human reference genome at a position where the Neanderthal genome looked like the chimpanzee genome, this did not necessarily mean that all humans today looked like the human reference genome. In fact, most such positions would vary among humans, but our knowledge about genetic variation among humans was too incomplete to differentiate real finds and false positives. Fortunately, there were several large projects under way aimed at describing the extent of genetic variation among humans, including the 1,000 Genomes Project, the goal of which was detecting all variants in the human genome present in 1 percent or more of humans. But that project was just starting. A third apparent limitation was that our genome was a composite of sequences from only three Neanderthals, and for most positions, we had only the sequence of a single Neanderthal individual. However, I didn't view this as overly problematic. As long as one single Neanderthal had the ape-like, ancestral version at a given position, it didn't matter if other Neanderthals that we hadn't sequenced carried the derived, new version that we saw in humans today. The knowledge that the ancestral variant was in at least one Neanderthal told us that it had still been around when Neanderthals and modern humans parted ways, perhaps 400,000 years ago. This made it a potential candidate for defining what might be universally modern human.

Janet and Martin compared the human reference genome with the chimpanzee, orangutan, and macaque genomes and identified all positions where they differed. They then compared all four genomes to our Neanderthal DNA sequences, being careful to compare only those Neanderthal DNA sequences for which we had complete certainty as to where they came from in the genome. They found that we had Neanderthal sequence coverage for 3,202,190 positions where nucleotide changes had occurred on the human lineage. For the vast majority of these positions, the Neanderthals looked like us, which was not surprising, given that we are much more closely related to Neanderthals than to apes. But for 12.1 percent of these positions, the Neanderthal looked like the apes. They then checked whether the ancestral variants seen in apes and Neanderthals were still present in some humans today; in most cases they found both the ancestral and the new variants in present-day humans. This was not surprising because the mutations responsible happened quite recently. But some of these new variants were, as far as we could tell, present in all humans today. These were the positions that we found particularly interesting.

Most tantalizing were those changes that might have functional consequences. First and foremost among these were the ones that change amino acids in proteins. Proteins, of course, are encoded by stretches of DNA sequence in the genome called "genes." Proteins are made up of strings of twenty different amino acids and perform many jobs in our bodies, such as regulating the activity of genes, building up tissues, and controlling our metabolism. As a result, changes in proteins are more likely to have consequences for an organism than a mutation randomly chosen from the set of all the mutations we identified. Such potentially meaningful mutations—which result in one amino acid in a protein being replaced by another one, or change how long a protein is—occur much less often during evolution than nucleotide substitutions that do not cause such dramatic alterations. Ultimately, Martin showed me a list of 78 amino-acid-altering nucleotide positions where, as far as we knew, all humans today were similar to one another but different from the Neanderthal genome and the apes. We expected to both add and subtract mutations from this list as both the Neanderthal genome and the 1,000 Genomes Project neared completion. So an educated guess might be that the total number of amino-acid changes that had spread to all modern humans since we separated from Neanderthals would be less than 200.

In the future, when we'll have a much fuller understanding of how each protein influences our bodies and minds, biologists will often be able to affix a function to a particular amino acid in a protein and to identify whether it functioned the same way in Neanderthals. Unfortunately, such a comprehensive knowledge of our genome and biology will likely be achieved only long after I have joined the Neanderthals in death. However, I take some solace in the thought that the Neanderthal genome (and the improved versions of it that we and others will achieve in the future) will be a crucial contribution to this endeavor.

For the moment, though, the 78 amino-acid positions provided us with very few and only the very crudest of insights. Just looking at what the changes were gave us very little idea about what might have changed in the biology of the first individual to carry the new variant. However, one thing we did notice was that there were five proteins that each carried not just one but two amino-acid differences. This was very unlikely to have occurred by chance if a total of 78 mutations were to be randomly scattered among the 20,000 proteins encoded by the genome. These five proteins may therefore have altered their functions recently in human history. It is even possible that they lost their function or importance so that they were now free to accumulate changes unhindered by any constraints imposed upon them by their function. Either way, we knew we had to take a closer look at these five proteins.

The first protein with two changes was involved in sperm motility. I was not very surprised by this. Among human and nonhuman primates alike, genes involved in male reproduction and sperm motility have been known to frequently change, probably due to direct competition between sperm cells from different males when females copulated with multiple partners. This overt competition means that any genetic change that makes a sperm cell more likely to fertilize the egg than its competitors, perhaps by swimming faster, will spread in the population. Such a change is considered to be under positive selection, because it increases the chance the individual with the mutation will leave progeny in the next generation. In fact, the more direct competition there is between sperm cells from different males in a single female (head-to-head, so to speak), the more positive selection can act. So there is a correlation between the level of promiscuity in a species and the extent to which positive selection can be detected in genes that have to do with male reproduction. Among chimpanzees, where a female in estrus tends to copulate with all males that happen to be available to her, there is more evidence for positive selection on such genes than among gorillas, where one dominant, silverback male tends to monopolize all females in his group. The sperm of a patriarchal gorilla silverback have all the time they need to fertilize the egg, since the sperm of younger and subordinate males cannot enter into the race. Or rather, the competition has already taken place at an earlier stage on the social level, when the hierarchy in the group was established. Amazingly, even crude measures such as the size of the testicles relative to the body reflect this difference in male competition for fertilizations. Whereas chimpanzees have large testicles, and the even more promiscuous but smaller bonobos carry around even more impressive sperm factories, the intimidatingly huge silverback gorillas have puny little testicles. Humans, as measured both by testicle size and evidence for positive selection on genes relevant for male reproduction, seem to be somewhere between the extremes of chimpanzee promiscuity and gorilla monogamy, suggesting that our ancestors may have been not so unlike us, vacillating between emotionally rewarding fidelity to a partner and sexually alluring alternatives.

The second protein on Martin's list that carried two changes had no known function—a reflection of our woefully inadequate knowledge of what genes do. A third one was involved in the synthesis of molecules necessary to produce proteins in the cells. I had no clue what that might mean, and wondered whether the gene actually had additional functions that were unknown to us—not at all an unlikely possibility given our limited knowledge about the function of genes. But the two remaining proteins with two amino-acid changes were both present in skin—one was involved in how cells attach to one another, particularly while wounds are healing, and the other was present in the upper layers of the skin, in certain sweat glands, and in hair roots. This suggested that something in the skin had changed during the course of recent human evolution. Perhaps future work will show that the former protein has something to do with the tendency for wounds to heal faster in apes than in humans, and that the latter has something to do with our lack of fur. But for the time being it is just not possible to tell. We are simply too ignorant about how genes affect the ways our bodies work.

A future version of Martin's and Janet's catalog, based on a complete version of the Neanderthal genome and more knowledge about genetic variation in people today, will contain positions in the human genome that changed between perhaps 400,000 years ago, when our ancestors parted ways with Neanderthals and then spread to become present in all humans, and about 50,000 years ago, when the "replacement crowd" fanned out across the globe, on the other. After that time, no further changes could be established in *all* humans simply because humans were spread out across continents. Based on the numbers we obtained using the parts of the Neanderthal genome we had, we estimated that the total number of DNA sequence positions at which the Neanderthals differed from all humans today will be on the order of 100,000. This will represent an essentially complete answer to the question of what makes modern humans "modern," at least from a genetic perspective. If in an imaginary experiment one were to change each of these 100,000 nucleotides back to their ancestral state in a modern human, the result would be an individual who, in a genetic sense, was similar to the common ancestor of Neanderthals and modern humans. In the future, one of the most important research objectives in anthropology will be to study this catalog in order to identify those genetic changes that are of relevance for how modern humans think and behave.

Chapter 21
Publishing the Genome

In science, very few results are definitive. In fact, soon after arriving at an insight, often after great effort, one can generally foresee imminent developments that will improve upon it. Yet at some point, it is necessary to draw a line and say that the time has come to publish. In the fall of 2009, I felt that we had reached that point.

The paper that we were going to write would be a milestone in several ways. Above all else, it was the first genome sequenced from an extinct form of humans. True, Eske Willerslev's group in Copenhagen, Denmark, had published a genome from a lock of Eskimo hair that spring. But the lock of hair was just 4,000 years old and had been preserved in the permafrost, and 80 percent of its DNA was human. The title of their paper said that they had sequenced an "extinct Palaeo-Eskimo," although I wondered what present-day Eskimos thought about the contention that they were extinct. The Neanderthals were truly old, truly extinct, a different form of humans, and of crucial evolutionary importance as the closest relative of all present-day humans, no matter where they live on the planet. I also felt that we had set the technical stage for much future work; unlike carcasses preserved in permafrost, the bones we had used hadn't been preserved in extraordinary ways. They were similar to thousands of human and animal bones found in caves in many parts of the world. I hoped that the techniques we had developed could now be used to recover whole genomes from many such remains. The finding most likely to create controversy was that Neanderthals had contributed parts of their genome to present-day people in Eurasia. But since we had come to this conclusion three times using three different approaches, I felt that we had definitively laid this question to rest. Future work would surely clarify the details of when, where, and how it had happened, but we had definitively shown that it *had* happened. The time had come for us to present our results to the world.

My ambition was to write a paper that would be as understandable as possible to a wide audience since not only geneticists would be interested in what we had done, but also archaeologists, paleontologists, and others. In fact, I was getting pressure from various directions to publish our findings. The *Science* editor was asking me when the paper would be submitted, and journalists kept calling not only me but other members of the team to ask when we would publish. I was starting to feel increasingly embarrassed about giving scientific talks that were focused more on technical issues than on what the genome told us, even though everybody realized that by now we must have interesting results to report. Despite the pressure, I felt that it was crucial to keep our main findings secret until publication. I worried that one of the fifty or so people in the know would tell a journalist that we had found evidence of Neanderthal gene flow in present-day people. If that happened, the news would quickly be all over the media.

An additional recurring worry was that another group would publish Neanderthal sequences before we did. This second worry was of course focused on one particular person: our previous partner and current competitor, Eddy Rubin at Berkeley, whom we knew had access to Neanderthal bones and the resources necessary to work on them. I thought about all the efforts expended by everyone involved in this project over the past four years and imagined what it would feel like to wake up to newspaper headlines saying that Neanderthals had contributed genes to people today, based on perhaps ten times less data than we had, analyzed in haste. Quite uncharacteristically, I even found myself fretting about this as I tried to fall asleep at night.

It was impossible to hide my worries during our weekly phone meetings. I started to reiterate that no one was allowed to say anything about any aspect of our results to the press, however pushy a journalist might be. That not a single consortium member ever did so is testimony to the loyalty of the entire team. I also started to pressure everyone in the consortium to deliver descriptions of what they had done. This was less easy for them to achieve. Some scientists are so driven by intellectual curiosity that once they've found the solution to a problem, they will be remiss in going through the tedium of writing it up and publishing it. This, of course, is very bad. Not only does the public, which has ultimately funded the research, have a right to learn about the results, but other scientists also need to know the details of how results were achieved so that they can improve and build on them. In fact, this is the main reason why, when scientists are being considered for appointments and promotions, they're judged not on how many interesting projects they have started but, instead, on how many projects they have finished and published. Some members of the consortium delivered their texts quickly, some slowly and in a preliminary form, and some not at all. I thought about how to pressure even distinguished colleagues to deliver their write-ups and finally came up with an idea: I needed to take advantage of their vanity.

Most scientists, like most people, want recognition for a job well done. They thrive on how often their papers are cited in other publications and how many invitations they get to deliver lectures. Apportioning the credit in our case would be difficult. Several groups and more than fifty scientists had contributed to our project and would appear as authors on the paper, and it would be hard to attribute credit to individuals for each of the different, often very creative and laborious analyses that had been done. In spite of this, everybody had worked selflessly toward the common goal, but it seemed only fair to apportion some individual credit. The question I faced was how to do that, and in the process also stimulate people to write quicker and well.

As is typical of many large scientific papers, most results presented in our paper would be presented as so-called supplementary material that wouldn't be included in the print journal but would instead be published electronically on the journal's website. The bulk of this considerable pile of material would be the technical minutiae interesting only to the experts. Normally, the authors of the supplementary material are the same and appear in the same order as on the paper. I decided to change that. I suggested that each section of this supplementary material would have separate authors and include a corresponding author to whom any interested readers would be referred in case they had questions. This system would make much clearer who had done which experiments and analyses. It would also make each person personally responsible for the quality of the section, as any glory—or any blame—would be directed at least partly to him or her. To further improve the quality, we assigned one member of the consortium not involved in that particular aspect of the work to carefully read such supplementary sections in order to find errors and faults in the presentation. This all helped. People actually delivered their supplementary sections, which eventually swelled to 19 chapters and 174 pages. My task became to modify these sections and write the main text that would be printed in the journal. In this, the ever energetic David Reich was a great help. There was much e-mailing about changes to the text of the main paper but finally, in the first days of February 2010, Ed Green submitted everything to *Science*.

On the first of March we received comments from three reviewers, and almost three weeks later we received comments from a fourth reviewer. It isn't unusual for reviewers to find many things to complain about in a manuscript. In this case, however, they didn't have much to say: the two years we spent trying to find flaws in each other's work had allowed us to find most of the weak points ourselves. Nevertheless, there was quite a lot of back and forth about the text with the editor. In the end, the paper appeared on May 7, 2010, complete with its 174 pages of supplementary material.[1] The paper was "more like a book than a scientific paper," as one paleontologist put it.

On the day the paper appeared, the two major institutions in the world that provide the scientific community with access to genome sequences, the European Bioinformatics Institute in Cambridge, England, and the Genome Browser maintained by the University of California at Santa Cruz in the United States, made the Neanderthal genome freely available to all. In addition, we made available in a public database all of the DNA fragments we had sequenced from the Neanderthal bones, including those that we had judged to be of bacterial origin. I wanted everyone to be able to check every detail of what we had done. And I wanted them to do a better job if they could.

With the appearance of the paper came the anticipated media frenzy. However, my previous dealings with journalists had left me somewhat jaded so I left Ed, David, Johannes, and the others in the consortium to deal with the press. In fact, the day our paper was published, I was scheduled to give a big lecture at Vanderbilt University in Nashville, Tennessee. That trip, which had been planned for a long time, was a convenient way for me to avoid the hype. But the excitement did rub off on my very friendly hosts in Nashville. When they learned that someone who sounded rather odd had called asking for me at my hotel, they worried about my safety, thinking of Christian fundamentalists who might be opposed to an evolutionary origin of humans. They had the police trace the phone call. It had come from the university campus; for some reason, this made them even more nervous so they had two police officers in civilian clothes follow me around everywhere I went on campus. This was the first time I have had bodyguards when giving a talk. I appreciated the concern for my safety, and the attention made me feel important. But those two huge men, in their dark suits and earpieces, eyeing everyone who approached me with suspicion, made the after-lecture mingling with faculty and students slightly awkward.

As it happened, the Neanderthal paper appeared the week before the 2010 Cold Spring Harbor Genome Meeting, so I went straight from Nashville to Long Island. I very much enjoyed presenting our main findings in the same auditorium where, four years earlier, I had announced our intention to do the project. I ended my talk by saying that I hoped the Neanderthal genome would prove a useful resource for scientists in the future. As it happened, the future came only five minutes after I had stepped down from the stage.

The speaker who followed me was Corey McLean, a graduate student from Stanford University. As I sat down, I vaguely thought to myself that I didn't envy him; following a talk that attracted a lot of attention was never easy. Very quickly I came to regret this condescending attitude. Corey gave a brilliant presentation. He had analyzed the genomes of humans and apes and identified a total of 583 large chunks of DNA lost in humans but present in the apes. He had then looked at what genes were in those regions and identified several interesting genes that had been lost in humans. One of these encoded a protein expressed in penile spines, which are structures on the penises of apes that cause males to ejaculate very quickly. These spines are not present in humans, which enables us to enjoy prolonged intercourse. The gene Corey had found to be lost might well be the reason for that. Another chunk that humans had lost encoded a protein that might limit the extent to which neurons divide and might have something to do with how the brain got larger in humans. This was fascinating! But what was most satisfying to me was that, in just the few days the Neanderthal genome had been publicly available, Corey had already checked the Neanderthal genome to see which of the deletions present-day humans share with Neanderthals. This was precisely how I had hoped our work would be applied, as a tool that would allow others to extend their own research by timing when changes had happened during human evolution. Corey had found that the Neanderthals did indeed have the penile spine deletion, so we immediately learned something about the intimate anatomy of Neanderthals that the fossil record couldn't tell us. The deletion involved in brain size was also shared with Neanderthals, a finding that we would have anticipated given our knowledge from fossils that their brains were as large as ours. But some of the other chunks that he hadn't yet investigated were not deleted in Neanderthals. Future work would show whether they truly were absent in all humans today and, if so, whether they had some likely consequences for how present-day humans differed from Neanderthals.

I couldn't find Corey after the session because of the many people wanting to talk to him as well as to me, but the next day I found him and told him how much I appreciated his work. I was so emotional about what he had done that I had to stop myself from hugging him. As far as I knew, he was the first person who had put our genome to use in his research.

This Neanderthal genome paper received far more reaction from the scientific community than any other paper I had published. Almost everyone was positive. The most positive comment came from John Hawks at the University of Wisconsin–Madison. A paleontologist trained by Milford Wolpoff, John is one of the architects of the multiregional hypothesis. He is quite influential in anthropology through his blog, where he thoughtfully and insightfully discusses new papers and ideas in anthropology. "These scientists have given an immense gift to humanity," he wrote on his blog. "The Neanderthal genome gives us a picture of ourselves, from the outside looking in. We can see, and now learn about, the essential genetic changes that make us human—the things that made our emergence as a global species possible. . . . This is what anthropology ought to be." Our group was pleased, of course. Only Ed tried to keep a cool distance: he e-mailed the entire consortium, saying, "Can somebody get John Hawks some oxygen?"

Only one solidly negative reaction came to my attention, from the well-known paleontologist Erik Trinkaus. Knowing that he tended to be negative about whether the study of genetics could make a real contribution to anthropology, I had sent him our paper a few days before publication to allow him some time to study it before journalists called to ask his opinion about it. I had hoped that reading our paper might convince him that we had done a good job, and we had even exchanged two e-mails in which I tried to resolve what I felt were his misunderstandings about what we had said in the paper. Given my efforts to reach out to Erik, I was disappointed when I got an e-mail from a journalist in Paris asking me for my reactions to excerpts from what must have been quite extensive comments that Erik Trinkaus had sent her about our paper. She quoted him as saying: "Briefly, we have had abundant fossil anatomical evidence of gene flow between Neanderthals and early modern humans, most likely as a result of Neanderthal populations being absorbed into those of expanding modern human ones around 40,000 years ago. In other words, the new DNA data and analysis [add] almost nothing new to the discussion. . . . Most of the authors of the new article are simply ignorant of that literature and do not understand the fossil data, living human diversity, or the behavioral/archeological context of the human evolutionary changes. . . . To sum up, the paper is the result of a very expensive, technologically complicated analysis that advances the study of modern human origins and of the Neanderthals very little and in some ways regresses it."

I was amazed that Erik could actually think we knew *less* after sequencing the Neanderthal genome than we had before. I ended up saying, "I am sad that Dr. Trinkaus thinks this adds so little to our knowledge of Neanderthals." Despite his reaction, I was confident that others would see that genetics and paleontology could complement each other.

There were many others who were interested in the Neanderthal genome—perhaps most surprisingly, some fundamentalist Christians in the United States. A few months after our paper appeared, I met Nicholas J. Matzke, a doctoral candidate at the Center for Theoretical Evolutionary Genomics at UC Berkeley. Unbeknownst to me and the other authors, our paper had apparently caused quite a flurry of discussion in the creationist community. Nick explained to me that creationists come in two varieties. First, there are "young-earth creationists," who believe that the earth, the heavens, and all life were created by direct acts of God sometime between 5,700 and 10,000 years ago. They tend to consider Neanderthals as "fully human," sometimes saying they were another, now extinct "race" that was scattered after the fall of the Tower of Babel. As a consequence, young-earth creationists had no problem with our finding that Neanderthals and modern humans had mixed. Then there are the "old-earth creationists," who accept that the earth is old but reject evolution by natural, nondivine means. One major old-earth ministry is "Reasons to Believe," headed by a Hugh Ross. He believes that modern humans were specially created around 50,000 years ago and that Neanderthals weren't humans, but animals. Ross and other old-earth creationists didn't like the finding that Neanderthals and modern humans had mixed. Nick sent me a transcript from a radio show in which he commented on our work, saying interbreeding was predictable "because the story of Genesis is early humanity getting into exceptionally wicked behavior practices," and that God may have had to "forcibly scatter humanity over the face of the Earth" to stop this kind of interbreeding, which he compared to "animal bestiality."

Clearly our paper was reaching a broader audience than we had ever imagined. But most people weren't shocked by the idea that their ancestors had interbred with Neanderthals. In fact, many seemed to find the idea intriguing—some, as had happened before, even volunteering to be examined for Neanderthal heritage. By early September, I started to notice a pattern: it was mostly men who wrote to me. I went back through my e-mails and found that forty-seven people had written to say they thought they were Neanderthals—and of these, forty-six were men! When I told my students, they suggested that perhaps men were more interested in genomic research than women. But that didn't seem to be the case, as twelve women had written to me not because they thought they were Neanderthals but because they thought their spouses were! Interestingly, not a single man had written to make such a claim about his wife (since then, however, one man has actually done so). I joked that some interesting genetic inheritance patterns were at work here that we needed to investigate. But what we were obviously seeing were the effects of the cultural ideas about what Neanderthals were like. In popular lore, Neanderthals are big, robust, muscular, somewhat crude, and perhaps a little simple. Some of these characteristics might be seen as acceptable or even positive in men, but they were clearly not conventionally seen as attractive in women. This idea was brought home to me when *Playboy* magazine called to ask for an interview about our work. I accepted, thinking that this would probably be my one and only chance to appear in *Playboy*. The magazine ended up writing a four-page story called "Neanderthal Love: Would You Sleep with This Woman?" The accompanying illustration was of a sturdy, very dirty woman wielding a spear on a snowy mountain ridge. That distinctly unattractive image probably explains why hardly any men volunteer the opinion that they are married to Neanderthals.

Another question that attracted a great deal of interest was what it might mean that people outside of Africa carried some Neanderthal DNA. Again, it was obvious that Neanderthals seemed to have a bad reputation. *Jeune Afrique,* a weekly news magazine that covers political and cultural issues in French-speaking Africa, set the tone by ending its story about our results with the following: "But one thing is . . . certain: given the apelike appearance of Neanderthals, those who still believe that sub-Saharan Africans are less advanced than white-skinned people do not understand anything."[2]

In general, I found that people's reactions to our work said more about their worldview than about anything that we could possibly know about what happened 30,000 or 40,000 years ago. For example, there were many who asked what the benefits could be of the Neanderthal pieces of DNA that had come over to people outside Africa. Although this might be a relevant question, it still made me wary because it seemed to imply that there must be something positive about these segments of DNA because they existed in

Europeans or Asians, who have often tended to regard themselves as superior to other populations. To me, the null hypothesis—that is, the baseline idea one starts out from when investigating a scientific issue—is always that a genetic change has no functional consequences whatsoever. One then tries to reject that hypothesis—for example, in this case, by studying patterns in how humans vary. Thus far, we had seen no hint of changes leading to a difference in function so my answer to those questions was that we had no reason to reject our null hypothesis. Maybe all we were seeing were the traces of the very natural act of intergroup relations in the distant past. Admittedly, we hadn't looked very hard yet. In fact, within a year of the publication of the Neanderthal genome, others found something.

Peter Parham is one of the world's foremost experts on the major histocompatibility complex (MHC), perhaps the most complicated genetic system in the human genome and the system on which I did my PhD work in Uppsala many years ago. The MHC encodes transplantation antigens, proteins that exist in almost all cells in our bodies. Their function is to bind fragments of proteins from viruses and other microbes that infect the cell and transport them to the cell surface, where they are recognized by immune cells. Such cells will then kill the infected cell, thus limiting the infection before it can spread throughout the body. The MHC was discovered not because of its normal function of fighting infections but because of the ferocious rejection reaction launched by the immune system against transplanted tissues such as skin, kidneys, or hearts. This rejection of transplanted tissues, which gave transplantation antigens their name, is possible because transplantation antigen proteins are extremely variable, encoded by MHC genes that exist in tens or even hundreds of different variants. So when a person receives a transplanted organ from an unrelated individual, the donor will always carry different transplantation antigen variants, and the recipient's immune system will therefore recognize the graft as foreign and attack it. Lifelong immunosuppressive treatment is needed to counter this reaction, even if the transplant comes from a relative of the recipient and is hence not too genetically dissimilar. In contrast, transplantations can be made between genetically identical twins with almost no immunological complications since they carry identical MHC genes and hence transplantation antigens. Why transplantation antigens are so variable is not yet fully understood, but it's probably because the presence of many different variants in individuals allows the immune system to better distinguish between infected and healthy cells.

Peter Parham looked at the fragments of Neanderthal DNA that mapped to the MHC genes encoding transplantation antigens, and Ed Green, who had by this point moved on to a faculty position at UC Santa Cruz, helped him to identify even more fragments that we had initially missed because of the unusual variability of these genes. A year after our paper appeared, they reported at a meeting that one particular MHC gene variant that is common in present-day Europeans and Asians but not yet seen in Africans was present in our Neanderthal genome. In fact, they claimed that about half of all copies of this gene carried by Europeans and 72 percent of those carried by people in China came from Neanderthals. Given that no more than 6 percent, at most, of the overall genomes of these people came from Neanderthals, this amazing increase in frequency of the MHC variants suggests that at least some helped the newly arrived replacement crowd survive. Peter suggested that because Neanderthals had already lived outside Africa for more than 200,000 years when first encountered by modern humans, their repertoire of MHC gene variants might have become adapted to fighting diseases local to Eurasia and perhaps absent in Africa. Thus, once a modern human received these genes from Neanderthals, this advantage drove the genes to high frequencies. In August 2011, Peter and his colleagues published a paper in *Science* describing these findings.[3]

On December 3, 2010, seven months after our paper appeared, I received an e-mail from Laura Zahn, the editor at *Science* who had handled our paper, with the news that it had been awarded the AAAS Newcomb Cleveland Prize. I had gotten a few scientific prizes in my career and they were always pleasant boosts for my self-confidence. But this was special to me. The Newcomb Cleveland Prize was established in 1923 and is awarded annually to the authors of the best research article or report published in *Science*. It had originally been called the $1,000 Prize, although by then it had been increased to $25,000. What pleased me most was that it was awarded to all the authors of the paper so it recognized what our consortium had collectively achieved. As Linda told me that night, "To publish a paper in *Science* is a big deal. But publishing the best paper in *Science* in a year? Most couldn't even dream of that."

I talked to David and Ed, the other two main authors of the paper, and we agreed to accept the prize together at the AAAS meeting in

Washington, DC, in February 2011. We decided to use the money to organize a meeting in Croatia, where the members of the consortium could meet in the fall of 2011 to discuss which direction the analysis of the Neanderthal genome should go in the future. I anticipated this experience to be a repeat of the intense Dubrovnik meeting in 2009. In fact, by the time I received the e-mail from Laura Zahn, we knew that we had more than just the Neanderthal genome to discuss at such a meeting. We had the genome of another extinct human, from another part of the world.

Chapter 22
A Very Unusual Finger

On December 3, 2009, I was attending a meeting on the rat genome at the Cold Spring Harbor Laboratory. I was there to describe a project on artificial domestication in rats that my group had been working on for the last few years. As I walked from the dining hall to the lecture hall after breakfast, my cell phone rang. It was Johannes Krause calling from Leipzig and he sounded strangely excited. I asked him what the matter was. He asked me if I was sitting down. When I said no, he said I'd better sit down before hearing what he had to tell me. Starting to worry that something terrible had happened, I sat down.

He asked me if I remembered a small bone that we had gotten from Anatoly Derevianko in Russia (see Figure 22.1). Anatoly is the president of the Siberian branch of the Russian Academy of Sciences and one of Russia's foremost archaeologists. He had started his career back in the 1960s and was not only very influential in the Russian academic world but also politically well-connected. Over the past several years that we had worked with him I had come to appreciate him more and more, also as a friend. Anatoly has a very warm smile and I have found him always polite and open to collaboration. He is also a very experienced field archaeologist and physically active. In the large lake close to his institute in Novosibirsk he was known to go for kilometer-long swims. Although very different in appearance from my professor of Egyptology, Rostislav Holthoer, who was of Russian descent, Anatoly shares with him the great capacity for friendship and loyalty I find typical of Russians. I count myself very fortunate to collaborate with him.

Some years previously, Anatoly had visited our laboratory and given us a few small bones in plastic bags. They had been excavated in a spot called Okladnikov Cave in the Altai Mountains in southern Siberia, where Russia, Kazakhstan, Mongolia, and China meet. These bones from Okladnikov Cave were too fragmentary to tell what type of human they had come

FIGURE 22.1. Anatoly Derevianko with colleagues. Photo: Bence Viola, MPI-EVA.

from, but we extracted DNA from them and showed that they contained Neanderthal mtDNA. Together with Anatoly, we then published a paper in *Nature* in 2007 that extended the range where Neanderthals had lived by at least 2,000 kilometers further east of what had been commonly believed.[1] Prior to our paper, no Neanderthal had been confirmed east of Uzbekistan.

In the spring of 2009 we received another bone fragment from Anatoly. His team had discovered that fragment during the previous year in Denisova Cave, another cave in the Altai region located in a valley that connects the Siberian steppes in the north to China and Mongolia in the south. The bone was minuscule, and I hadn't attached very much importance to it, thinking only that we would see whether it contained any DNA at some point in the future when there was time. Perhaps it would prove to be Neanderthal, which would enable us to gauge the extent of mtDNA variation among the easternmost Neanderthals.

Johannes had now found the time to extract DNA from the bone; and Qiaomei Fu, a talented young graduate student from China, had made a library and used a method that Adrian Briggs, the British graduate student in our lab, had developed to fish out mtDNA fragments from the library. They found a very large amount of mtDNA—in total, 30,443 fragments, which enabled them to assemble the complete mitochondrial genome with a very high degree of accuracy. In fact, each position in the mtDNA was seen an average of 156 times, unusually high for an old bone. That was good news, but it wasn't why Johannes asked me to sit down. He had compared the mtDNA sequence of the Denisova bone to the six complete Neanderthal mtDNA sequences that we had previously determined as well as to mtDNA sequences from present-day humans from around the world. Whereas the Neanderthals differed from modern humans at an average of 202 nucleotide positions, the Denisova individual differed at an average of 385 positions—almost twice as much! In a tree analysis, the Denisova mtDNA lineage branched off well before the modern human and Neanderthal lineages shared a common ancestor. When Johannes calibrated the rate of substitutions by assuming that humans and chimpanzees split 6 million years ago, then the Neanderthal mtDNA split from the human lineage about half a million years ago—just as we had previously shown—and the mtDNA of the Denisova bone branched off approximately 1 million years ago! I could hardly believe what Johannes was telling me. This was neither a modern human nor a Neanderthal! It was something else, entirely.

My head was spinning. What extinct human group could have split off from the human lineage a million years ago? *Homo erectus?* But the oldest *H. erectus* fossils outside Africa were found in Georgia and were about 1.9 million years old. So *H. erectus* were supposed to have left Africa and thus to have split from the lineage leading to present-day humans almost 2 million years ago. *Homo heidelbergensis?* But they were thought to be the direct ancestors of Neanderthals and would then presumably have diverged from the modern human lineage at the same time as Neanderthals. Was this bone from something totally unknown? A new form of extinct human? I asked Johannes to tell me everything about this bone.

The bone was indeed tiny, the size of two grains of rice put together. It came from the last phalanx of the little finger (see Figure 22.2), the outermost part of a pinky, from what was probably a young individual. Johannes had used a dentistry drill to remove thirty milligrams of material from the bone, and from this tiny amount of bone powder he had extracted the DNA that Qiaomei had used to make the library. Given how much mtDNA she and Johannes found, the DNA preservation in the bone must be exceptionally good. I would be back in Leipzig in three days and I told him that we would meet then and decide what to do.

After I hung up, I couldn't bring myself to listen to presentations about how the genomes of different rat strains differed from each other. It was a sunny and snowless winter day in the New York area. I spent the

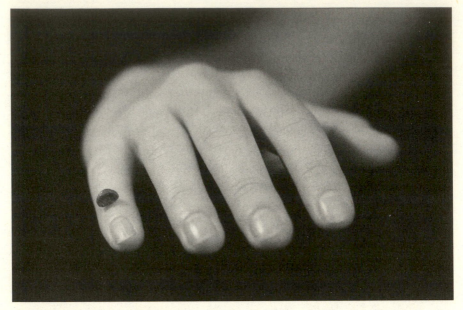

FIGURE 22.2. The small finger bone discovered in 2008 by Anatoly Derevianko and Michael Shunkov in Denisova Cave. Photo: MPI-EVA.

morning walking along the windy beach below Cold Spring Harbor and
thought about the young person who had died far away in a Siberian cave
many thousands of years ago. All that remained of that life was a tiny speck
of bone, but it was enough to tell us that she represented something unknown to us, a group of humans who had left Africa before the ancestors
of the Neanderthals but after *Homo erectus*. Could we find out what this
group was?

When I got back to Leipzig I sat down with Johannes and the others to discuss our next steps. The analyses of the Neanderthal genome were drawing
to a close, so people had time on their hands to think about these startling findings. The first thought was whether there could be something
wrong with the DNA sequence Johannes had reconstructed. Qiaomei and
Johannes had retrieved thousands of mtDNA fragments and much less
than 1 percent of them carried substitutions that were suggestive of contamination. Since the mtDNA looked quite different from present-day human mtDNA, it couldn't be contamination from anyone today. Earlier in
my career I had often worried about fragments of mtDNA that thousands
or millions of years in the past had become integrated on a cell's nuclear

chromosomes. Such mtDNA fossils could sometimes be mistaken for an actual mtDNA sequence. Fortunately, the circular shape of the mtDNA enables us to separate nuclear mtDNA fossils from real ones. The DNA sequence Johannes had reconstructed from overlapping fragments must have derived from a circular molecule. I didn't see how our findings could be wrong. Nevertheless, Johannes and Qiaomei would do a separate, independent DNA extraction from some powder that was left of the bone sample and repeat what they had done. But this was more of a formality. I was certain he would get the same results.

I turned to the question of what this unusual person could have been. If there were more bones in the cave, that could help us figure this out. I was told that Anatoly Derevianko had given us only a portion of the bone so there must be a bigger piece still in Novosibirsk. Perhaps there were also other bones that would give us a hint of what this person could have looked like or that we could use to extract more DNA. We clearly needed to visit Novosibirsk.

I immediately e-mailed Anatoly and said that we had some very unexpected and exciting results and that I wanted to present these to him in person as soon as possible. I said that we would also be very interested in further analyzing the other part of the bone, and perhaps to date it. Anatoly answered the next day and asked for more details about the results. I summarized them for him and we arranged for me to visit Novosibirsk, along with Johannes as well as Bence Viola, a jovial archaeologist of Hungarian descent, in mid-January 2010. Bence specializes in Central Asian and Siberian paleontology, and we had often worked with him in the past. I had just succeeded in convincing him to come to Leipzig from Vienna to work with us and the paleontologists at our institute. A fourth person who would join us was Victor Wiebe. He had done his PhD in Novosibirsk in the '70s and knew Anatoly and several other people there from that time, and he had been working with me for twelve years. On this trip he would serve as a much-needed interpreter. I had studied Russian thirty-five years earlier, during my military service in Sweden, but by that point I remembered only crude questions one would pose to prisoners of war. They were not suited for scientific discussions.

After a stopover in Moscow and a long overnight flight on to Novosibirsk, we landed in the early morning of January 17. A digital display at the airport terminal showed the time as 6:35 a.m. Then it switched to the temperature: −41°C. When our luggage arrived I opened my bag and put on all the clothes I had. The air outside the terminal was very dry and the snow was like powder swirling around our feet as we quickly made our way to the car. When I breathed in, the sides of my nose tended to freeze to the septum.

The drive to Akademgorodok took an hour. As the name suggests, Akademgorodok is a city that was built solely for scientific pursuits in the 1950s by the Soviet Academy of Sciences. At its heyday, it housed more than 65,000 scientists and their families. After the collapse of the Soviet Union, many scientists left Akademgorodok and most institutes there had declined. But by 2010, the Russian governments and several large companies had been investing money in the city for almost ten years, and there was a sense of tentative and cautious optimism around the city.

We were housed at the Golden Valley Hotel, which had been converted from a typical Soviet nine-story apartment building. I had visited the hotel once before; one of my most vivid memories from that visit was the lack of hot water, which drove me every morning to walk a good half-hour through the birch forests to swim in a nearby reservoir called the Ob Sea. That, however, was in summer, and I was more than a bit concerned about whether the heating system would work now. I needn't have worried. When Johannes and I arrived at our room, we found not only was there hot water in the tap but the radiators were so hot that the temperature in the room was unbearably warm—about 40°C. There was no valve to turn the heating down, so we ended up opening the windows and letting in outside air that was almost 80° colder. We kept that window open for the duration of our stay.

We arrived on a Sunday, and our meeting with Anatoly wasn't until the next day, so after a nap the four of us decided to take a walk. We were amazed to find a small ice cream vendor open for business. Feeling certain that this was the one and only time I would ever have an ice cream when it was −35°C out, I approached the shack. The woman who sold me the ice cream realized that I wasn't local and urged me to eat the ice cream fast: once it reached ambient temperature, it would be rock-hard and impossible to eat. After quickly eating the ice cream we walked through the frosty forest to the beach where I had swum during warm summer mornings two years earlier. We were the only ones there. The sky was clear but the pale sun provided not even a trace of warmth. Fortunately there was no wind. Even the tiniest bit of air that made it into our clothing had a chilling effect. In fact, by this point my toes were numb and we quickly retreated to our overheated hotel rooms.

The next day we met with Anatoly in the spacious office he enjoyed as head of the Institute of Archaeology and Ethnography. Michael Shunkov, the archaeologist who led the excavations at Denisova Cave, was also present, as were some of their associates. Johannes presented his and Qiaomei's findings, and everyone was taken aback. Was this a new form of extinct human, perhaps some form that was present only in Siberia or only in the Altai Mountains? There were in fact several plant and animal species endogenous to the Altai area, so the idea was certainly plausible. Over a lunch of delicious Russian cold cuts washed down with vodka, all served in Anatoly's office, we excitedly discussed what we might have found. After some time, when the atmosphere was both animated and relaxed, I pointed out that the ultimate answer to our questions would be found in the nuclear genome. If we could sample the remaining, larger piece of the finger bone we would be able to sequence the nuclear genome and get a more complete picture of how this individual was related to people today and to the Neanderthals whose genomes we had just sequenced. At first I didn't understand Anatoly's answer to this request, which I blamed on my bad Russian and inebriated condition. But I was still puzzled after Victor's translation. Anatoly was apparently explaining that he no longer had the other piece of the bone since he had given it to my "friend" about a year ago. Bewildered, I looked questioningly at Victor, Bence, and Johannes. What friend of mine? Did one of them already have it? But they looked as stunned as I was. Then Anatoly clarified. He had given it to "my friend Eddy, Eddy Rubin, at Berkeley."

I have no idea how I looked or what I said after that. I knew that Eddy had been trying to get his hands on bones to sequence the Neanderthal genome ahead of us. But here we were learning that, for almost a year, he had had a much larger piece than we did of this particular bone that contained so much endogenous DNA that it would be possible to sequence the nuclear genome in a matter of a few weeks, without technical tricks or many hundreds of runs on sequencing machines. And we were still weeks from even submitting our Neanderthal paper to *Science,* let alone its publication. My recurring worst worry suddenly seemed about to come true: before we could publish there would be a paper from Berkeley presenting the genome of another extinct form of humans, sequenced to even higher coverage than the Neanderthal genome. Who would then care about our years spent painstakingly working on extraction techniques, on enriching for endogenous DNA, on teasing out the Neanderthal DNA from the vast excess of bacterial DNA? All these details would be important in the long run for use on the hundreds of bones that weren't as miraculously well preserved as this one, but in terms of getting the genome of an extinct human relative, Eddy would have done it faster and better, simply because he got lucky.

I struggled to regain my composure and to say something that wouldn't give my feelings away. But I managed only some mumblings about scientific collaboration. We soon left the meeting with a plan to meet our hosts for dinner later at the House of Scientists, the social center of Akademgorodok. Walking back to the hotel room, I didn't feel the cold anymore. Johannes tried to console me. He tried to make me see that we should just continue doing the best work we could and forget about the competition. He was right, of course. But obviously we shouldn't drag our feet. Now, more than ever, we had to be fast.

Dinner was an ebulliently friendly affair, as were all the dinners I have enjoyed with Anatoly. The food was excellent: salmon, herring, and caviar were followed by several delicious main courses. Toasts with good vodka were frequent throughout the evening and, as is customary in Russia, each dinner participant took his turn proposing a toast to some commonly appreciated theme, such as collaboration, peace, our teachers, our students, love, women, and so on. When I had first started traveling in the Soviet Union, I had loathed this custom, feeling immensely embarrassed when I had to mumble my way through a speech on a theme I didn't enjoy talking about in front of a large dinner crowd. With time, though, I had gotten used to it and had even come to appreciate the fact that it allowed all participants at the dinner, even those whose social standing would not normally have allowed them to be heard, let alone dominate the conversation, to command everyone's undivided attention for a short time.

Undoubtedly I also had come to appreciate this custom thanks to the fact that, deep down, I'm a very sentimental person, a trait that alcohol often helps bring to the surface. And sentiments are what these toasts are about. I toasted, first, to our very fruitful collaboration and then to peace, pointing out how I had grown up in capitalist Sweden, and had been conditioned to regard a huge war in Europe as a likely scenario and Russia as our natural enemy. Since Sweden was officially neutral, the potential enemy I had been trained to face during my military service was officially and euphemistically named "the superpower," but tellingly, the language spoken with prisoners during our war games was Russian. But the war everyone planned for never came. We never had to face each other as enemies. Instead we were sitting here as friends, working together and discovering amazing things together. Thanks to the alcohol, I was moved by my own words. As one of the youngest at the dinner, Johannes appropriately chose to toast his teachers. I realized how inebriated I was when he brought tears to my eyes by saying that he had two fathers in science: me, who had introduced him to molecular evolution and ancient DNA, and Anatoly Derevianko, who, during two field trips to the Altai and Uzbekistan, had introduced him to archaeology. In fact, I was so moved because these were truths that we wouldn't normally share with each other.

We walked to our hotel after dinner along the main street of Akademgorodok. The night was very cold and dark, and the stars unbelievably bright due to the fact that the ice-cold air could hold hardly any humidity. But I didn't notice. The tension from earlier in the day had caused me to down shots of vodka faster than I normally would have. In fact, I had the feeling that I hadn't been so drunk since my teens. But as we unsteadily made our way down the snowy street, Bence told me something that instantly penetrated even my intoxicated mind. Earlier in the visit, Anatoly had given him a tooth that had been found nine years earlier in Denisova Cave. It was a molar (see Figure 22.3), probably from a juvenile, but it was huge. Bence said that he had never before seen a similar tooth, seeming unlike both Neanderthal and modern human teeth. In fact, he said, if he hadn't known where it had been found, he would have thought that it had come from some much older human ancestor, maybe *Homo erectus* in Africa, or *Homo habilis,* or maybe even *Australopithecus.* It was the most amazing tooth he had ever seen. In our drunken state, we were sure that it must have come from the same person as the finger bone, and we felt certain this creature must really have been something we hadn't seen before. In the Altai, there have long been rumors of mountain-dwelling snow men called Almas. As we made our way toward the hotel, we shouted that we had found an Alma! We joked that, if we could get a radiocarbon date from the tooth, we might find it to be just a few years old. This would explain why it contained so much DNA. Maybe these Yeti-like creatures were still living somewhere on the border between Russian and Mongolia. I don't quite remember finding our hotel room and getting into bed that night.

The next morning it was difficult to get up to catch the taxi to the airport, and none of us spoke much until an hour or two into our flight to Moscow. By then, the bleak reality of our situation was slowly dawning on me, tainted by the dreariness and cold sweat of a serious hangover. Maybe they were already writing a paper in Berkeley on the Denisova bone. We had started writing a paper on our Denisova mtDNA results over Christmas, but it was now urgent that we finish this paper as soon as possible. Where

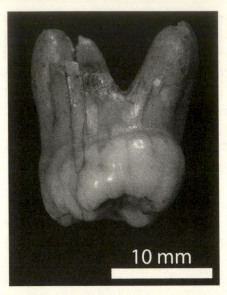

FIGURE 22.3. The Denisova molar. Photo: B. Viola, MPI-EVA.

would we submit it? The editors at *Science* were already impatiently waiting for our Neanderthal genome paper. To approach them about a different paper on a different topic might make them write us off as unable to finish one project, let alone two. So we decided to contact *Nature*. During a long layover at the airport in Moscow, I wrote an e-mail to Henry Gee, the senior editor who handles paleontology at *Nature,* and to Magdalena Skipper, the editor who handles genomics. I told them that we had a paper almost finished that described "what we interpret as a new hominin species based on a complete mitochondrial DNA sequence that diverged from the human line about twice as long ago as the Neanderthal mtDNA." I was all too aware that the publication process could drag on for many months. It could even end in rejection, after months of dithering with reviewers and editors, after which we would need to submit to another journal and endure another similarly lengthy process. I didn't want that to happen this time so I told them that we had direct competition and would be grateful if the paper could be handled quickly. An hour and fifteen minutes later, Henry Gee replied with "How exciting! Prediction is very hard, especially about the future. However, when you send it in, we'll give it topmost priority."

As soon as we were back in Leipzig we finished the manuscript, which we entitled "The Complete mtDNA Genome of an Unknown Hominin from Southern Siberia," and sent it off to *Nature*. It was a unique paper. For the

first time ever, a new form of extinct humans was described from DNA sequence data alone, in the total absence of any skeletal remains. Given that
the mtDNA was so different from that of both modern humans and Neanderthals, we felt sure that we had found a new form of extinct human. In
fact, we were so taken with this idea that, after some discussion, we decided
to describe it as a new species, which we called *Homo altaiensis.*

However, I felt vaguely uneasy about suggesting a new species and
soon had second thoughts. To me, taxonomy, the classification of living
organisms into species, genera, orders, and so on, is a sterile academic exercise, particularly when discussing extinct human forms. Whenever my
students send me manuscripts in which they use Linnaean Latin names for
groups that are commonly known—for example, "In order to better understand the pattern of genetic variation in *Pan troglodytes,* we sequenced . . . "
—I always delete the Latin and sometimes even snidely ask who they are
trying to impress by saying "*Pan troglodytes*" instead of "chimpanzees."
Another reason I dislike taxonomy is that it has a tendency to elicit scientific debates that have no resolution. For example, if researchers refer
to Neanderthals as "*Homo neanderthalensis,*" they indicate that they regard them as a separate species, distinct from "*Homo sapiens.*" This invariably infuriates multiregionalists, who see continuity from Neanderthals
to present-day Europeans. If researchers say, "*Homo sapiens neanderthalensis,*" they indicate that they see them as a subspecies, on par with
"*Homo sapiens sapiens.*" This invariably infuriates proponents of the strict
out-of-Africa hypothesis. These arguments I prefer to avoid, and although
we had by now shown (but not yet published) that there had been mixing
between Neanderthals and modern humans, I knew that taxonomic wars
over Neanderthal classification would continue, since there is no definition
of a species perfectly describing the case. Many would say that a species
is a group of organisms that can produce fertile offspring with each other
and cannot do so with members of other groups. From that perspective
we had shown that Neanderthals and modern humans were the same species. However, this concept has its limitations. For example, polar bears
and grizzlies can (and occasionally do) produce fertile offspring with each
other when they meet in the wild. Yet polar bears and grizzlies look and
behave differently, and are adapted to different lifestyles and environments.
It would seem rather arbitrary, if not outright ridiculous, to regard them as
one and the same species. We didn't know whether the fact that Neanderthals contributed perhaps 2 to 4 percent of the genes of many present-day
humans meant that they were the same or different species. So it was ironic
that, having always refrained from using a Latin name for Neanderthals in

our papers, I was now on the verge of introducing a new Linnaean species designation myself.

Despite my misgivings about fruitless taxonomic debates, I felt I had some reasons for this digression from my principles. The mtDNA of the Denisova individual was about twice as different from the mtDNAs of modern humans as was the mtDNA of Neanderthals. That probably made them more like *H. heidelbergensis,* who did get to have their own Latin species name. But there was also vanity involved. Not many people get to name a new hominin species, which made it tempting to do so, even more so because this was the first time it would be done based solely on DNA data. However, the deciding argument came both from some people in our group and from Henry Gee at *Nature.* He pointed out that if we didn't take the initiative and give this hominin group a species name, someone else would. And that person might come up with a name we didn't like. So, after deliberating with Anatoly and the team who had excavated the crucial finger bone, we settled on provisionally naming it *Homo altaiensis.*

Nature kept its promise to process our paper quickly: eleven days after our submission, we received comments from four anonymous reviewers. They all praised the technical aspects of the paper but they were divided on the issue of naming a new species. Two reviewers voiced concerns that we might actually have sequenced a late *Homo erectus.* They felt that if *H. erectus* had had continuous contact with groups in Africa, they may not show an mtDNA divergence as deep as their first exit out of Africa some 2 million years ago. I doubted this. But the fourth reviewer made the point that saved us from ourselves. He or she said that "once a name is in the taxonomic literature, it cannot be withdrawn later. So such provisional naming is not wise, I believe." When I read this, I realized we had been foolish.

In the meantime, it dawned on us that the very large amounts of mtDNA that Johannes had been able to capture from the Denisova DNA libraries meant we would be able to sequence quite a bit of this individual's nuclear genome. This would settle its relationships both to Neanderthals and to modern humans in a definitive way as well as its possible status as a new species. We rewrote the manuscript and removed any reference to a new species. Instead, we said that "nuclear DNA sequences are needed to clarify definitively the relationship of the Denisova individual to present-day humans and Neanderthals." We sent it back to *Nature,* where it appeared in early April.[2] As events would show, we had reason to be grateful that we had not named it a new species.

Chapter 23
A Neanderthal Relative

We began the nuclear DNA sequencing from the libraries that Johannes had prepared from the bone as soon as we could. The results were stunning. When Udo mapped them to the human genome, he found matches for about 70 percent of all DNA fragments. Yet contamination with modern human DNA, as judged from the mtDNA results, was extremely low. This meant that more than two-thirds of the DNA in the bone had come from the dead individual! By comparison, only 4 percent of the DNA from our very best Neanderthal remains did so; more typically, the proportion was well below 1 percent. This bone was as well-preserved as the mammoth that Hendrik Poinar had sequenced and the Eskimo that Eske Willerslev in Copenhagen had sequenced. But both of those specimens had been deep frozen in the permafrost shortly after death. This explained why the majority of the DNA in those specimens was not bacterial, but I could not explain why the individual from Denisova Cave had produced so much DNA. Whatever the reason, it certainly made the analysis of the genome much easier. In fact, our biggest issue was how to weed out the microbial DNA fragments in the library rather than how to fish out the few endogenous DNA fragments, as we had done with the Neanderthals. Now the major question was a good one: Just how much of the nuclear genome could we get? As always, we didn't want to use the outermost surface of the bone fragment. First, it seemed irresponsible to use it all up, as we didn't know how much of the larger piece Eddy and his group had used up in Berkeley. Second, if any part of the bone was contaminated from people handling it, it would be the surface. So Johannes used the internal part of the bone to produce two extracts. From test runs of libraries prepared from these DNA extracts, Martin Kircher calculated that we would be able to get even more coverage of the genome than we had for the Neanderthals.

When Johannes made libraries from the extracts, he applied one of Adrian Briggs's innovations to deal with the chemical damage that changed C nucleotides in the DNA to U nucleotides. Adrian had shown that most of these U nucleotides were found close to the ends of the ancient DNA molecules, and how to remove the damaged ends. In doing so, he lost an average of one or two nucleotides at the ends of about half the ancient molecules but he also got rid of the vast majority of errors in the DNA sequences. Since it was no longer necessary to take frequent C to T errors into account, the mapping of the fragments to the human genome became easier. Johannes made two large libraries with this method. Not only were about 70 percent of the DNA fragments in those libraries from the Denisova individual, but those DNA fragments now carried many fewer errors than the Neanderthal DNA fragments. This was real progress. Yet, I was nervous, knowing that Eddy's group might also be at work on the same project, or even polishing a nice manuscript presenting the genome. So I tried to get everything to move as fast as possible, asking the sequencing groups to set other projects aside and sequence these libraries as fast as they could.

I was also very curious about the strange-looking tooth that Anatoly had given us. Only DNA work would tell us if it came from the same type of person as the finger bone. Johannes, as careful as any dentist treating a live patient, drilled a small hole in the tooth and made extracts from the powder he retrieved and, in turn, libraries from the DNA in the extracts. From the libraries he then fished out mtDNA fragments. In addition, we immediately sequenced random DNA fragments from the libraries to see how much of the DNA was endogenous to the individual.

There was good news and bad news. The good news was that he was able to reconstruct the entire mtDNA genome. There were two differences between it and the finger bone, which meant both that it was from a different person and that they were the same type of humans. The bad news was that the fraction of endogenous DNA in the tooth was only 0.2 percent. We were now even more mystified about why the finger bone contained so much endogenous DNA. I speculated that the finger might have rapidly desiccated after death, which might have limited the degradation of the DNA by enzymes in the dying cells and stopped bacterial growth. I joked that perhaps this person died with her pinky pointing up into the air so that it mummified before bacteria had too much of a chance to multiply.

Now that we had shown that the tooth came from the same type of human as the finger, Bence devoted himself to the analysis of its morphology with renewed energy. Although I am no tooth expert, even I found it to be startlingly large. It was almost 50 percent larger than my molars. Bence pointed out that besides being very big, it was different from most Neanderthal molars with respect to both the absence and presence of certain traits in its crown. Also, its roots were unusual. Unlike Neanderthal molar roots, which tend to be closely spaced or even fused, it had strongly diverging roots. Bence concluded that the tooth morphology suggested that the Denisova population was distinct from both Neanderthals and modern humans. In fact, since the Denisova tooth lacked Neanderthal features that evolved about 300,000 years ago, he surmised that the ancestors of Denisova individuals had gone their separate way from Neanderthals before that. This was in line with what the mtDNA told us. But I was always cautious, some might even say overly skeptical, about the interpretation of morphological traits. Perhaps the Denisova people had reverted to having ancient-looking teeth after separating from either modern humans or Neanderthals. Only the nuclear genome would tell the complete story.

Our sequencing machines began churning out Denisova nuclear DNA sequences at around the same time that we were dealing with the reviewers' comments and finalizing the Neanderthal paper. Thus, we didn't have much time to look at the Denisova sequences immediately, but I imagined that we could analyze them quickly once we got to it. During the last four years we had developed computer programs to analyze the Neanderthal genome that could now be directly applied to the genome from the Denisova individual. Still, I remained afraid that Eddy might be far ahead of us, so I decided to scale down the Neanderthal Genome Analysis Consortium to a core and, I hoped, faster group, asking them to devote their full attention to the Denisova genome. Most crucially, we needed David Reich, Nick Patterson, and Monty Slatkin and his crew (see Figure 23.1). We initially called ourselves the "X-Man" group because we didn't know what the Denisova individual was. Bence had by then told us that the finger was from a young individual, perhaps just three to five years old, and we had sequenced the maternally inherited mtDNA so it seemed inappropriate to use a designation that made everyone think of a macho comic figure. I considered "X-Girl" but thought that sounded too much like a Japanese manga character. Finally, I settled on "X-Woman"—and the name stuck. Right away, the X-Woman Consortium began having weekly phone meetings.

Udo mapped the DNA fragments to the human and chimpanzee genomes. It was comparatively easy given that we had used Adrian's approach to remove the majority of the errors, but Udo warned me that the

FIGURE 23.1. Monty Slatkin, Anatoly Derevianko, and David Reich, at a meeting at Denisova Cave in 2011. Photo: B. Viola, MPI-EVA.

mapping was preliminary. In spite of this, we distributed the data to the X-Woman Consortium. Not long after we had submitted the final version of the revised mtDNA paper to *Nature,* Nick Patterson sent me a report on its preliminary analysis of Udo's preliminary mappings. When I read it, I felt grateful to the reviewer who had convinced us not to name a new species. Nick had found two things.

First, he found that the nuclear genome of the Denisova finger bone was more closely related to the Neanderthal genome than to the genomes of people living today. In fact, it seemed to be only slightly more different from the Neanderthal genome than the deepest differences one could find among humans living today—for example, between the Papua New Guinean individual we had sequenced and the African San individual. This was quite a different picture than the one painted by the mtDNA results alone, and my immediate suspicion was that gene flow from some other more ancient hominin in Asia was responsible for introducing the mtDNA into the Denisova individuals. After all, we had just shown that modern humans had interbred with Neanderthals, so gene flow seemed a reasonable guess. But it was something we needed to think carefully about.

The second thing Nick had found was even more unexpected. Among the five humans we had sequenced for the Neanderthal analysis, the Denisova individual shared more derived SNP alleles with the Papuan individual than with the Chinese, European, or two African individuals. One possible explanation was that relatives of the Denisova individual had mixed with the ancestors of the Papuan individual, although given the distance from Siberia to Papua New Guinea I felt we might be jumping to conclusions. There could be some systematic error in what we did, and Udo again warned me that his mappings of the DNA fragments to the genome were preliminary. Perhaps there was something in the complex computer analyses that created extra similarity both between the Denisova and Neanderthal genomes and between the Denisova and Papuan genomes. Then both of Nick's findings could be wrong.

A week later Ed finished his own careful analysis of the new data. He found that there were very few Y chromosomal fragments among the DNA we had sequenced, so X-Woman really was a woman, or rather, given the tiny bone, a girl. The general lack of Y chromosomal fragments also indicated that male nuclear DNA contamination was low. When he looked at divergence of the Denisova DNA sequences from the human and Neanderthal genomes, he, like Nick, found that the Denisova genome shared more derived SNP alleles with the Neanderthal genome than with modern humans. So this suggested that the common ancestor of the Denisova girl and Neanderthals first diverged from the lineage that includes modern humans, and only then did the ancestor of the Denisova girl and Neanderthals go different ways. In other words, the Denisova girl and Neanderthals were more closely related to each other than they were to modern humans. Several questions arose as we discussed these data during our Friday meetings in Leipzig and during long phone meetings with Nick, David, Monty, and the others. How could the Denisova mtDNA be so different when the Denisova nuclear genome was closer to Neanderthals than to modern humans? Could the Denisova girl perhaps have had recent ancestors who included Neanderthals and some more archaic human form, perhaps late *Homo erectus*? Or could she be a mixture of modern humans and such an archaic hominin? We looked at each of these possibilities and none seemed to fit.

It took Udo a few months to refine the mapping of all the fragments to each of the comparison genomes. The final mappings didn't change the picture, and I became convinced that the Denisova girl was a member of a population that shared a common origin with Neanderthals, but that had lived separately from the Neanderthals for at least as long as Finns today have been separated from, say, the San in southern Africa. Denisova DNA sequences tended to be a bit closer to those of Eurasians than to Africans, but less so than were the Neanderthal DNA sequences. This was best explained by a common ancestry for the Denisova girl and Neanderthals so that when Neanderthals mixed with modern humans, Eurasian ancestors inherited DNA sequences that were somewhat similar to Denisova DNA sequences just because the Neanderthals were related to the Denisova girl.

So it was clear that the population to which the Denisova girl belonged had separated from Neanderthals before they met modern humans. What would we call this population? We certainly didn't want to give them a Latin name that would force us to label them a subspecies or a species. Since they were only about as different from the Neanderthals as I am from a San, this would be ridiculous. But we needed to call them something. We needed what taxonomists would call a trivial name, such as "Finn," "San," "German," or "Chinese." "Neanderthal" was such a trivial name, named for Neander Valley in Germany, *Thal* being an old spelling of the German word for "valley." Following this example, I suggested that we call them "Denisovans." Anatoly agreed, so we unceremoniously announced our decision in a phone meeting and, from then on, we referred to the population that included X-Woman and the individual with the unusually large molar as the Denisovans.

One exciting issue remained: whether Nick's finding, that the Denisovan girl shared more derived sequence variants (SNPs) with the Papuan individual than with the other four individuals we had sequenced, was a real discovery or instead due to some bug in a computer program or a quirk in the data. Over the next several weeks, we discussed different technical problems that could cause that data to look this way. But things remained ambiguous. There could, perhaps, be something special about the Papuan DNA sequences that made them appear to be slightly more similar to the Denisovan DNA sequences. To me, it seemed suspicious that we had seen no trace of this putative admixture in China since it would mean that Papuan ancestors could have met the Denisovans, who we knew existed in Siberia, without meeting the ancestors of the Chinese. Of course, maybe the Denisovans lived in other places beyond Siberia. We decided that the best way to address this was to sequence more present-day people. This slowed down our progress toward publication, but we didn't want to make fools of ourselves by claiming something that would then turn out to be due to some technical oversight on our part. So we decided to sequence seven more people from around the world. We chose an African Mbuti and a European from Sardinia, two people we wouldn't expect to have anything to do with the Denisovans. We also included a person from Mongolia in central Asia as a person who lived not too far from the Altai area; a Cambodian as someone on mainland Asia not too far from Papua; and a Karitiana from South America as a representative of Native Americans, whose ancestors had come from Asia and could perhaps have met Denisovans in the past. Finally, we decided to sequence two people from Melanesia and chose a second Papuan and a person from the island of Bougainville.

With those sequences in hand, Nick and the others redid their analyses. Their results confirmed that the Denisovan genome had a special relationship with the people from Papua and Bougainville. In contrast, there was no extra sharing of derived SNPs whatsoever with the people from Cambodia, Mongolia, or South America.

Martin also found another interesting thing. He detected an indication that the Denisovan genome carried slightly more ancestral (ape-like) sequence variants than the Neanderthal genome. This could indicate gene flow into Denisovan ancestors from some archaic human that could also have brought in the diverged mtDNA. But both Nick and Monty were still worried that we might be overlooking some artifact. Could it be risky to do detailed analyses with the Neanderthal and Denisovan genomes together? Since they were both ancient genomes they might share some errors that were the result of being deposited in the soil for thousands of years. There was even some discussion about whether the gene flow into Papuans might after all be due to some esoteric technical problem.

By the end of May I was growing increasingly frustrated. After a long phone meeting with what seemed to me to be unnecessarily complicated discussions about possible technical problems, I wrote an e-mail to the consortium in a fit of bad temper, saying that I felt that our major contributions to the scientific community were the Denisovan genome sequence itself as well as the Denisovan tooth with its unusual morphology. So far, the world knew only about the Denisova mtDNA sequence and therefore thought that modern humans and Neanderthals were each other's closest relatives and that the Denisova individual was a more distant relative. From the nuclear genome, we now knew that the real situation was that Denisovans and Neanderthals were closer to each other and that modern humans were their more distant relatives. We needed to tell this to the world as soon as possible and let other researchers have access to the genome we had sequenced. If we weren't sure of whether or not there had been admixture with the Papuans, we simply didn't need to discuss that issue in the paper. It could be addressed in a later paper when we had time to explore it more fully.

This was a deliberately provocative suggestion, and many smart people in the consortium were opposed to it. Adrian wrote an e-mail saying "Surely publishing without the Papuan story risks the following: Someone will do their own analysis, find the Papuan admixture story, and publish it quickly. Why we didn't mention it ourselves will then be interpreted as a) incompetence, b) rushing, c) political correctness. Isn't that a problem?" Nick agreed, saying "We have to deal with the Papuan issue or we will look like fools or cowards."

So we continued to struggle with figuring out what technical issues might have caused this unexpected result. What finally turned the tide was that Nick analyzed the relationship of the Denisovan genome to another, publicly available data set. The Human Diversity Panel, available from a center in Paris, is a collection of cell lines and DNA from 938 humans from 53 populations from around the world. Each sample had been analyzed by a "gold standard" technology that shows with great accuracy which nucleotide is present at 642,690 variable sites in the genome. Nick looked at how often the Neanderthal and Denisovan genomes shared derived SNPs at places where we had good data for both ancient genomes. He found that all seventeen individuals from Papua New Guinea and all ten individuals from Bougainville stood out from all other individuals outside Africa in that they were closer to the Denisovan genome. This was in perfect agreement with what we had found when we analyzed the genomes we had sequenced. We were all now convinced that something special had indeed gone on between the Denisovans and the ancestors of the Papuans.

Using the Denisova and Neanderthal genome data, David and Nick estimated that about 2.5 percent of the genomes of people outside Africa came from Neanderthals, and that later gene flow had brought about 4.8 percent of Denisovan DNA into the Papuans. Since Papuans also carried the Neanderthal component in their genome, this meant that approximately 7 percent of the genomes of Papuans came from earlier forms of humans. This was an amazing finding. We had studied two genomes from extinct human forms. In both cases we had found some gene flow into modern humans. Thus, low levels of mixing with earlier humans seemed to have been the rule rather than the exception when modern humans spread across the world. This meant that neither Neanderthals nor Denisovans were totally extinct. A little bit of them lived on in people today. It also meant that Denisovans must have been widespread in the past, although it is curious that they don't seem to have mixed with modern humans in Mongolia, China, Cambodia, or anywhere else on mainland Asia. A plausible explanation was that we had found the traces of admixture between the first modern humans to have migrated out of Africa, moving along the southern coast of Asia, before the rest of Asia was populated by modern humans. Many paleontologists and anthropologists have speculated about such an early coastal migration of modern humans from the Middle East to southern India, the Andaman Islands, Melanesia, and Australia. If these people met and mixed with Denisovans, perhaps in present-day Indonesia, then their descendants in Papua New Guinea and Bougainville, and presumably also Australian Aborigines, would all carry Denisovan DNA. Maybe we didn't see evidence of admixture with Denisovans elsewhere in Asia because other modern human groups that later colonized mainland Asia followed more inland routes and so never mixed with Denisovans. Or maybe they didn't even meet because Denisovans were already extinct by the time they arrived.

Later on, after our paper describing the Denisovan genome had appeared, Mark Stoneking in our department together with David performed a much more detailed genetic survey of Southeast Asian populations and found Denisovan admixture in Melanesia, Polynesia, and Australia and in some populations in the Philippines, but not on the Andaman Islands, and nowhere else in the region. Thus, the idea that the early modern migrants out of Africa who came along a southern route met Denisovans and mixed with them somewhere on mainland Southeast Asia seems a plausible one.

Monty Slatkin used all the DNA sequences we had generated to test various population models. As I expected, he found that the simplest model that explained all the data was admixture between Neanderthals and modern humans, followed by later admixture between Denisovans and Melanesian ancestors. But we still needed to explain the very strange Denisovan mtDNA. There were two possibilities. One was that the mtDNA lineage was introduced into Denisovan ancestors through admixture with another, more archaic hominin group. This was the idea I secretly favored. The other was that it was due to a process known as "incomplete lineage sorting." This means simply that the population that was the common ancestor of Denisovans and Neanderthals as well as modern humans carried earlier versions of all three mtDNAs. Then, by chance, one mtDNA variant that carried a lot of differences from the other two became the one that survived in Denisovans whereas the other two, which were much more similar to each other, became the ones that survived in Neanderthals and modern humans, respectively. This was particularly likely to have occurred if the ancestral population of Denisovans, Neanderthals, and modern humans was large enough that many mtDNA lineages could have coexisted in it. Monty's population models showed that the data could be explained either by a small amount of admixture from another unknown human group or by this "incomplete lineage sorting" scenario. Although that meant we couldn't favor one explanation over another, admixture nonetheless seemed a more plausible explanation to me. After all, we had already detected two cases of mixture between archaic groups and modern humans, so I had become much more open to the possibility that mixing was a common feature during human evolution. Furthermore, if the Denisovans were willing to have sex with modern humans, it seemed plausible that they would have sex with other archaic groups as well. I had come to believe that although the big picture of modern human spread was one where the replacement crowd pushed other groups into extinction, this was not a total replacement. Rather, some DNA seemed to leak over into the groups that lived on, so much so that I started using a term I had picked up from somewhere to describe this process: "leaky replacement." Perhaps, I thought, the spread of Denisovans had also been a "leaky" affair.

In July, we started writing the paper. Since 70 percent of the DNA in the Denisovan bone was endogenous, the sequencing of the Denisovan genome was much less of a *tour de force* than sequencing the Neanderthal genome had been, but that meant we had been able to produce a better-quality genome sequence, with slightly higher coverage (1.9-fold instead of 1.3-fold) for the Denisovan sample. But more importantly, the removal of the deaminated C's had reduced the numbers of errors in it so that they were about five times less frequent than in the Neanderthal genome. We submitted the paper to *Nature* in the middle of August. I felt it was an amazing paper. From a bone about one-quarter the size of a sugar cube, we had determined a genome sequence and used it to demonstrate that it came from a previously unknown human group. It showed that molecular biology could contribute fundamentally new and unexpected knowledge to paleontology.

Nature again sent our paper out to four anonymous reviewers. The comments we received differed in quality, from the jealously quarrelsome to the insightfully critical. As with our earlier mtDNA paper, one of the reviewer's comments ended up substantially improving our paper. He or she pointed out potential problems with the analyses where we had used the

Neanderthal and Denisovan genomes together to suggest that archaic gene flow was likely to have contributed the mtDNA to the Denisovans. I felt that we had dealt with those problems adequately, but the reviewer made us take the safer route and avoid such analyses entirely. His or her review also made us do more work to show that the signals of gene flow into Melanesians could not be due to differences in DNA preservation, sequencing technology, or other differences in how the data were collected. When we resubmitted the paper after taking the comments into account, this reviewer graciously acknowledged our efforts, saying "often, when one raises concerns regarding the underlying analytical methodology used to arrive at a conclusion, . . . the concerns are explained away by the authors. . . . Here, the authors have done the opposite: they have taken my comments very seriously, investigated the issues I raised, and undertaken a substantive revision of their work to address my concerns." I felt like a schoolboy being praised by his teacher. The reviewer even identified himself: it was Carlos Bustamante, a population geneticist at Stanford whom I had always respected.

In late November 2010, *Nature* accepted our paper for publication. The editor suggested that we delay publication until mid-January in order to get more press coverage and attention than would be possible during the Christmas holidays. We discussed this in the consortium. Some agreed with the editor. I felt that if we had worked as fast as we could in view of the potential competition, then we shouldn't delay the last step. Against what was perhaps a majority opinion I pushed for publication as soon as possible and the paper ended up appearing on December 23.[1] I'm sure this caused it to get less attention that it would have otherwise, but I felt good about the fact that it came out the same year as the Neanderthal genome.

When Linda, Rune, and I drove up to our small house in snowy Sweden that Christmas, I felt that it had truly been an exceptional year. We had achieved even more than I had dreamt we would. But even though we had sequenced the Neanderthal genome and opened the door to the genomes of other extinct human groups, many mysteries remained. One big mystery was when the Denisovans had lived. Both the finger bone fragment and the tooth were too small to allow us to obtain radiocarbon dates. Instead, we had dated seven bone fragments, most with cut marks or other human modifications, found in the same layer in Denisova Cave. Four of the seven turned out to be older than 50,000 years, while three were between 16,000 and 30,000 years old. So it seemed there had been humans in the cave before 50,000 years ago and then again after 30,000 years ago. I tended to think that the older people were the Denisovans and the younger people modern humans, but we couldn't be sure. Professor Shunkov and Anatoly had found amazingly sophisticated stone tools and a polished stone brace-let in what seemed to be the same layer as the finger bone. Could they have been made by the Denisovans? It was an outlandish idea but the archaeol-ogists felt it was possible.

Another big mystery was how far the Denisovans had ranged. We knew that they were in southern Siberia, but the fact that they had met and conceived children with the ancestors of Melanesians suggested that they had been much more widespread in the past. Perhaps they had roamed all over Southeast Asia, from temperate or even subarctic regions to the trop-ics. I thought we needed to look for Denisovan DNA in fossils from China. It would also be extremely exciting if Anatoly and his team could find more complete remains of Denisovans in the Altai Mountains. If those bones had features that set Denisovans apart from other hominin groups, these features would perhaps allow us to identify other fossils elsewhere in Asia as Denisovans.

My group and others have since gone on to work on these mysteries. Still other groups have begun to use ancient DNA to study past human ep-idemics and prehistoric civilizations. But that December I felt a satisfaction rare in my scientific career. What started as a secret hobby when I was a graduate student in my native Sweden over thirty years ago had resulted in a project that seemed like science fiction when we announced it a little over four years earlier. We had now brought this project to a successful conclu-sion. With my family in our cozy little Swedish hut, I was more relaxed over those Christmas holidays than I had been for a long time.

Postscript

Three years later, as I write this, we still do not know what happened to the other part of the finger bone that Anatoly sent to Berkeley. Perhaps one day it can be used for dating so that we will know when the Denisova girl lived.

Anatoly and his team have continued to unearth amazing bones in Denisova Cave. They have found another huge molar that contains Denisovan DNA. They have also found a toe bone that turned out to come from a Neanderthal.

David Reich and his postdoc Sriram Sankararaman have used genetic models to date the admixture between Neanderthals and modern humans to sometime between 40,000 and 90,000 years ago.[1] This shows that actual interbreeding between Neanderthals and modern humans has caused the extra similarity between the Neanderthal genome and the genomes of people in Europe and Asia, not the more complicated scenario of ancient substructure in Africa that we also considered in 2010.

Matthias Meyer, something of a technical wizard in our lab, has developed new and amazingly sensitive methods to extract DNA and make libraries. This has allowed us to use the tiny leftover fragments of the Denisova finger bone to sequence its genome to a total coverage of 30-fold.[2] Recently, we have followed up by sequencing the Neanderthal genome from the toe bone found in Denisova Cave to 50-fold coverage. These ancient genomes are now of higher accuracy than most genomes determined from people living today.

When we compare the Neanderthal genome to the genome of the Denisovan girl, we see that she carried a component in her genome from a hominin that diverged from the human lineage earlier than Neanderthals and Denisovans. We also see that Denisovans mixed with Neanderthals, and that they contributed small amounts of DNA not only to people in Melanesia but also to people who live on mainland Asia today. These were subtle signals of past mixing that we could not see in 2010, when we worked with genomes of lower quality. The picture that emerges is that there was plenty of mixing among several types of humans in the late Pleistocene, but mostly of small proportions.

Together with new data from the 1,000 Genomes Project, these two archaic genomes of high quality now allow us to create a near-complete catalog of sites in the genome where all people today are different from Neanderthals and Denisovans as well as from the apes. This catalog contains 31,389 single nucleotide changes and 125 insertions and deletions of a few nucleotides. Of these, 96 change amino acids in proteins, and perhaps 3,000 affect sequences that regulate how genes are turned on and off. There are surely some nucleotide differences, particularly in repetitive parts of the genome, that we have missed, but it is clear that the genetic "recipe" for making a modern human is not very long. The next big challenge is to find out what the consequences of these changes are.

George Church, a brilliant technical innovator at Harvard University, has suggested that scientists should use our catalog to modify a human cell back to the ancestral state and then use that cell to recreate or "clone" a Neanderthal. In fact, already when we announced that we had completed the Neanderthal genome sequencing at the AAAS meeting in 2009, George was quoted by the *New York Times* as saying that "a Neanderthal could be brought to life with present technology for about $30 million." He added that if someone were eager to supply the financing, he "might go along with it." To his credit, he acknowledged that there are ethical problems with such a project, but suggested that to avoid those, one could use not a human cell but a chimpanzee cell!

This, as well as later statements to the same effect, I write off as George's tendency to be provocative. Nevertheless, they point to a dilemma. How do we study traits specific to humans—for example, language or aspects of intelligence—when for both technical and ethical reasons we cannot do what George suggests? The way forward is, on the one hand, the introduction of human and Neanderthal genetic variants into the genomes of human and apes cells that can then be used not to clone individuals but to study their physiology in a plastic dish in the laboratory and, on the other hand, the introduction of such variants into laboratory mice. Our laboratory in Leipzig has already taken the first steps in that direction. In 2002, we found that the protein made from a gene called *FOXP2*, which Tony Monaco's group in Oxford, England, had shown to be involved in language ability in humans, differed at two amino-acid positions from the same protein in apes and almost all other mammals.[3] Encouraged by the fact that the mouse *FOXP2* protein is very similar to the *FOXP2* protein of the chimpanzee, we decided to introduce the two human changes into the mouse genome. It took several years of hard work by a talented student, then postdoc, then group leader in our lab, Wolfgang Enard, until the first mice that made the human version of the *FOXP2* protein were born. The results greatly exceeded my expectations. The peeps the pups produced at about two weeks of age when removed from the nest differed subtly but significantly from those of their non-humanized littermates, supporting the idea that these changes have something to do with vocal communication. This finding has led to much more work showing that the two changes affect how neurons extend outgrowths to contact other neurons and how they process signals in parts of the brain that have to do with motor learning.[4] At the moment, we are collaborating with George Church to put these changes into human cells that can be differentiated to neurons in the test tube.

Although the two changes in *FOXP2* are actually shared with Neanderthals and Denisovans,[5] these experiments nevertheless point to how, in the future, we may sort out which changes are crucial for what makes modern humans special. One can imagine putting such changes into cell lines, and into mice, alone and in different combinations, in order to "humanize" and "neanderthalize" biochemical pathways or intracellular structures, and then to study their effects. One day, we may then be able to understand what set the replacement crowd apart from their archaic contemporaries, and why, of all the primates, modern humans spread to all corners of the world and reshaped, both intentionally and unintentionally, the environment on a global scale. I am convinced that parts of the answers to this question, perhaps the greatest one in human history, lies hidden in the ancient genomes we have sequenced.

Notes

Chapter 1

1. R. L. Cann, Mark Stoneking, and Allan C. Wilson, "Mitochondrial DNA and human evolution," *Nature* 325, 31–36 (1987).

2. M. Krings et al., "Neandertal DNA sequences and the origin of modern humans," *Cell* 90, 19–30 (1997).

Chapter 2

1. S. Pääbo, Über den Nachweis von DNA in altägyptischen Mumien," *Das Altertum* 30, 213–218 (1984).

1. S. Pääbo, "Preservation of DNA in ancient Egyptian mummies," *Journal of Archaeological Sciences* 12, 411–417 (1985).

Chapter 3

1. S. Pääbo, "Molecular cloning of ancient Egyptian mummy DNA," *Nature* 314, 644–645 (1985).

2. S. Pääbo and A. C. Wilson, "Polymerase chain reaction reveals cloning artefacts," *Nature* 334, 387–388 (1988).

3. R. L. Cann, Mark Stoneking, and A. C. Wilson, "Mitochondrial DNA and human evolution," *Nature* 325, 31–36 (1987).

4. W. K. Thomas, S. Pääbo, and F. X. Villablanca, "Spatial and temporal continuity of kangaroo-rat populations shown by sequencing mitochondrial-DNA from museum specimens," *Journal of Molecular Evolution* 31, 101–112 (1990).

5. J. M. Diamond, "Old dead rats are valuable," *Nature* 347, 334–335 (1990).

6. S. Pääbo, J. A. Gifford, and A. C. Wilson, "Mitochondrial-DNA sequences from a 7,000-year-old brain," *Nucleic Acids Research* 16, 9775–9787 (1988).

7. R. H. Thomas et al., "DNA phylogeny of the extinct marsupial wolf," *Nature* 340, 465–467 (1989).

8. S. Pääbo, "Ancient DNA—Extraction, characterization, molecular-cloning, and enzymatic amplification," *Proceedings of the National Academy of Sciences USA* 86, 1939–1943 (1989).

Chapter 4

1. S. Pääbo, R. G. Higuchi, and A. C. Wilson, "Ancient DNA and the polymerase chain reaction," *Journal of Biological Chemistry* 264, 9709–9712 (1989).

2. G. Del Pozzo and J. Guardiola, "Mummy DNA fragment identified," *Nature* 339, 431–432 (1989).

3. S. Pääbo, R. G. Higuchi, and A. C. Wilson, "Ancient DNA and the polymerase chain reaction," *Journal of Biological Chemistry* 264, 9709–9712 (1989).

4. T. Lindahl, "Recovery of antediluvian DNA," *Nature* 365, 700 (1993).

5. E. Hagelberg and J. B. Clegg, "Isolation and characterization of DNA from archaeological bone," *Proceedings of the Royal Society B* 244:1309, 45–50 (1991).

6. M. Höss and S. Pääbo, "DNA extraction from Pleistocene bones by a silica-based purification method," *Nucleic Acids Research* 21:16, 3913–3914 (1993).

7. M. Höss and S. Pääbo, "Mammoth DNA sequences," *Nature* 370, 333 (1994); Erika Hagelberg et al., "DNA from ancient mammoth bones," *Nature* 370, 333–334 (1994).

8. M. Höss et al., "Excrement analysis by PCR," *Nature* 359, 199 (1992).

9. E. M. Golenberg et al., "Chloroplast DNA sequence from a Miocene Magnolia species," *Nature* 344, 656–658 (1990).

10. S. Pääbo and A. C. Wilson, "Miocene DNA sequences—a dream come true?" *Current Biology* 1, 45–46 (1991).

11. A. Sidow et al., "Bacterial DNA in Clarkia fossils," *Philosophical Transactions of the Royal Society B* 333, 429–433 (1991).

12. R. DeSalle et al., "DNA sequences from a fossil termite in Oligo-Miocene amber and their phylogenetic implications," *Science* 257, 1933–1936 (1992).

13. R. J. Cano et al., "Enzymatic amplification and nucleotide sequencing of DNA from 120–135-million-year-old weevil," *Nature* 363, 536–538 (1993).

14. H. N. Poinar et al., "DNA from an extinct plant," *Nature* 363, 677 (1993).

15. T. Lindahl, "Instability and decay of the primary structure of DNA," *Nature* 362, 709–715 (1993).

16. S. R. Woodward, N. J. Weyand, and M. Bunnell, "DNA sequence from Cretaceous Period bone fragments," *Science* 266, 1229–1232 (1994).

17. H. Zischler et al., "Detecting dinosaur DNA," *Science* 268, 1192–1193 (1995).

Chapter 5

1. H. Prichard, *Through the Heart of Patagonia* (New York: D. Appleton and Company, 1902).

2. M. Höss et al., "Molecular phylogeny of the extinct ground sloth *Mylodon darwinii*," *Proceedings of the National Academy of Sciences USA* 93, 181–185 (1996).

3. O. Handt et al., "Molecular genetic analyses of the Tyrolean Ice Man," *Science* 264, 1775–1778 (1994).

4. O. Handt et al., "The retrieval of ancient human DNA sequences," *American Journal of Human Genetics* 59:2, 368–376 (1996).

5. In fact, even at this writing, several groups are using the PCR to study mtDNA from human archaeological remains without describing clearly how they distinguish contaminating DNA sequences from endogenous ones. Some of the sequences they determine are almost certainly correct, but others are almost equally certainly incorrect.

Chapter 6

1. I. V. Ovchinnikov et al., "Molecular analysis of Neanderthal DNA from the northern Caucasus," *Nature* 404, 490–493 (2000).

2. M. Krings et al., "A view of Neandertal genetic diversity," *Nature Genetics* 26, 144–146 (2000).

Chapter 8

1. H. Kaessmann et al., "DNA sequence variation in a non-coding region of low recombination on the human X chromosome," *Nature Genetics* 22, 78–81 (1999); H. Kaessmann, V. Wiebe, and S. Pääbo, "Extensive nuclear DNA sequence diversity among chimpanzees," *Science* 286, 1159–1162 (1999); H. Kaessmann et al., "Great ape DNA sequences reveal a reduced diversity and an expansion in humans," *Nature Genetics* 27, 155–156 (2001).

2. D. Serre et al., "No evidence of Neandertal mtDNA contribution to early modern humans," *PLoS Biology* 2, 313–217 (2004).

3. M. Currat and L. Excoffier, "Modern humans did not admix with Neandertals during their range expansion into Europe," *PLoS Biology* 2, 2264–2274 (2004).

Chapter 9

1. A. D. Greenwood et al., "Nuclear DNA sequences from Late Pleistocene megafauna," *Molecular Biology and Evolution* 16, 1466–1473 (1999).

Chapter 10

1. H. N. Poinar et al., "Molecular coproscopy: Dung and diet of the extinct ground sloth *Nothrotheriops shastensis*," *Science* 281, 402–406 (1998).

2. S. Vasan et al., "An agent cleaving glucose-derived protein cross-links in vitro and in vivo," *Nature* 382, 275–278 (1996).

3. H. Poinar et al., "Nuclear gene sequences from a Late Pleistocene sloth coprolite," *Current Biology* 13, 1150–1152 (2003).

4. J. P. Noonan et al., "Genomic sequencing of Pleistocene cave bears," *Science* 309, 597–600 (2005).

5. M. Stiller et al., "Patterns of nucleotide misincorporations during enzymatic

amplification and direct large-scale sequencing of ancient DNA," *Proceedings of the National Academy of Sciences USA* 103, 13578–13584 (2006).

6. H. Poinar et al., "Metagenomics to paleogenomics: Large-scale sequencing of mammoth DNA," *Science* 311, 392–394 (2006).

7. See note 5 above.

Chapter 11

1. J. P. Noonan et al., "Sequencing and analysis of Neandertal genomic DNA," *Science* 314, 1113–1118 (2006); R. E. Green et al., "Analysis of one million base pairs of Neanderthal DNA," *Nature* 444, 330–336 (2006).

Chapter 12

1. After our *Nature* publication, we learned that it should more appropriately be called Vi-33.16, according to a more recent numbering system.

2. R. W. Schmitz et al., "The Neandertal type site revisited: Interdisciplinary investigations of skeletal remains from the Neander Valley, Germany," *Proceedings of the National Academy of Sciences USA* 99, 13342–13347 (2002).

3. A. W. Briggs et al., "Patterns of damage in genomic DNA sequences from a Neandertal," *Proceedings of the National Academy of Sciences USA* 104, 14616–14621 (2007).

Chapter 13

1. T. Maricic and Svante Pääbo, "Optimization of 454 sequencing library preparation from small amounts of DNA permits sequence determination of both DNA strands," *BioTechniques* 46, 5157 (2009).

2. J. D. Wall and Sung K. Kim, "Inconsistencies in Neandertal genomic DNA sequences," *PLoS Genetics* 10:175 (2007).

3. A. W. Briggs et al., "Patterns of damage in genomic DNA sequences from a Neandertal," *Proceedings of the National Academy of Sciences USA* 104, 14616–14621 (2007).

Chapter 14

1. R. E. Green et al., "The Neandertal genome and ancient DNA authenticity," *EMBO Journal* 28, 2494–2503 (2009).

Chapter 15

1. R. E. Green et al., "A complete Neandertal mitochondrial genome sequence determined by high-throughput sequencing," *Cell* 134, 416–426 (2008).

Chapter 16

1. N. Patterson et al., "Genetic evidence for complex speciation of humans and chimpanzees," *Nature* 441, 1103–1108 (2006).

Chapter 20

1. M. Tomasello, *Origins of Human Communication* (Cambridge, MA: MIT Press).

Chapter 21

1. R. E. Green et al., "A draft sequence of the Neandertal genome," *Science* 328, 710–722 (2010).

2. My translation.

3. L. Abi-Rached et al., "The shaping of modern human immune systems by multiregional admixture with archaic humans," *Science* 334, 89–94 (2011).

Chapter 22

1. J. Krause et al., "Neanderthals in central Asia and Siberia," *Nature* 449, 902–904 (2007).

2. J. Krause et al., "The complete mtDNA of an unknown hominin from Southern Siberia," *Nature* 464, 894–897 (2010).

Chapter 23

1. D. Reich et al., "Genetic history of an archaic hominin group from Denisova Cave in Siberia," *Nature* 468, 1053–1060 (2010).

Postscript

1. S. Sankararaman et al., "The date of interbreeding between Neandertals and modern humans," *PLoS Genetics* 8:1002947 (2012).

2. M. Meyer, "A high coverage genome sequence from an archaic Denisovan individual," *Science* 338, 222–226 (2012).

3. W. Enard, et al., "Molecular evolution of *FOXP2*, a gene involved in speech and language," *Nature* 418, 869–872 (2002).

4. W. Enard et al. "A humanized version of *Foxp2* affects cortico-basal ganglia circuits in mice," *Cell* 137, 961–971 (2009).

5. J. Krause et al., "The derived *FOXP2* variant of modern humans was shared with Neandertals," *Current Biology* 17, 1908–1912 (2007).

Index